A Practical Guide to Cost Engineering

A Practical Guide to Cost Engineering aims to show you how to work as a cost engineer out in the real world.

Written by an experienced cost engineer and training program developer, this book introduces the practical side of cost management (cost estimation, cost reduction, and cost control) through real cases and realistic examples from a diverse range of engineering-based projects. With examples from nuclear, oil and gas, and renewable energy sectors, the book introduces and demonstrates the activities of the cost engineer throughout a project life cycle. The content is divided into logical sections covering basic concepts, cost estimation, cost control, economic feasibility, sustainability, and more, and the chapters are packed full of features such as definitions, formulas, exercises, and examples. The focus is on providing a practical approach where the reader can first understand a concept and then apply it using an Excel tool developed by the author which allows the reader to simulate different scenarios and results.

The simple approach focusing on essential information backed up by practical scenarios presented in this book allows cost engineers and related professionals to execute and understand their activities, develop their professional skills, and even develop in-house training programs.

Helber Macedo is a Costing Manager at Baker Hughes working with Subsea Projects in the oil and gas industry in the UK. He has worked as a cost engineer in the oil and gas industry for over 15 years. He is the author of a number of technical papers and developed and taught a variety of in-house cost estimating, construction, planning, and industrial assembly courses for Petrobras.

A Practical Guide to Cost Engineering

Helber Macedo

Routledge
Taylor & Francis Group

LONDON AND NEW YORK

Cover image: © Getty Images

First published 2024
by Routledge
4 Park Square, Milton Park, Abingdon, Oxon OX14 4RN

and by Routledge
605 Third Avenue, New York, NY 10158

Routledge is an imprint of the Taylor & Francis Group, an informa business

British Library Cataloguing-in-Publication Data
A catalogue record for this book is available from the British Library

ISBN: 978-1-032-51535-9 (hbk)
ISBN: 978-1-032-50582-4 (pbk)
ISBN: 978-1-003-40272-5 (ebk)

DOI: 10.1201/9781003402725

Typeset in Times
by Deanta Global Publishing Services, Chennai, India

To Juliana, Julia, and Fernanda

"What no eye has seen and no ear has heard, what the mind of man cannot visualise; all that God has prepared for those who love him."

I Cor. 2:9

Contents

Preface

It has been said that engineers are good at calculating but not at writing. This statement doesn't apply to Helber Macedo, whose commendable book *A Practical Guide to Cost Engineering*, I had the honor (and pleasure) to read, learn from, underline, and share with my colleagues.

Macedo's guide spans subjects such as cost estimate, contingency, management reserves, economic indicators, economic feasibility study, cost control, and statistics in a clear and straightforward manner, with plenty of examples and hints, and a unique practical approach which differentiates this book from some arid and boresome publications. This book represents Macedo's bright career shedding light on grey zones of cost engineering, helping clarify concepts, and introducing international best practices.

The guide is most certainly useful to beginners to the art/science of cost engineering, to mature engineers who can brush up their skills and knowledge, and to senior managers who need to be acquainted with the general tools and techniques of project controls. If I could edit this book, I would add an expression to the end of the first paragraph: "Project Management has many challenges because of increasing market competitiveness, risks, uncertainties, *and poor knowledge of the practitioners.*" It is exactly to improve the overall quality of the practitioners that the guide accomplishes its meritorious mission.

Aldo D. Mattos
Fellow of AACE International

Acknowledgments

Thanks to all my friends who collaborated in my career as a cost engineer and in the book revision, particularly Jose R. Paiva, Carlos E. Braga, Aldo Mattos, Marcos H. Silveira, Yadu Poudel, and Kul B. Uppal.

1 Introduction

Project management has many challenges because of increasing market competitiveness, risks, and uncertainties. One is to avoid cost overruns, which means executing the project by spending more than planned. In this scenario, cost engineering techniques should be applied to ensure the best performance.

Cost engineering is the science of cost management along the project lifecycle. It covers a range of methodologies related to cost estimate, cost control, cost forecast, economic studies, planning and scheduling, and risk management.

Throughout my career as a cost engineer, I detected a gap in books and references in this area. This book aims to fill it through a practical approach where an industry vision is adopted to support or clarify the learning path of initial or experienced professionals.

The book covers the cost engineering subject divided into three main sections:

- Cost estimation
- Cost control
- Economic feasibility study

Also, additional sections cover sustainability, statistics and maths, and basic concepts throughout the book. Hence, the book aims to show the cost engineering activities during the project lifecycle.

Definitions, formulas, examples, and exercises sequence each chapter. The focus is on providing a hands-on, clear, and straightforward style where the reader can understand the concept and how to apply it.

Chapter 2 explains concepts used throughout the book, such as cost, price, direct and indirect cost, overhead, CAPEX, OPEX, scope, work breakdown structure, cost breakdown structure, fixed, semi, and variable cost, and the tripe labour, material, and equipment.

Chapter 3 shows cost estimate methods for the planning phase. Methods covered include physical dimensions, end-product units, capacity factor, factor method, and the parametric equation. They are called quick methods because it is possible to get a cost estimate through a few minutes and inputs. The examples are based on renewables and O&G industries.

Chapter 4 explains the deterministic cost estimate or bottom-up estimate. It discusses the planning activities that support the cost estimate, the basis of estimate (BOE), cost review, cost validation, and accuracy range and finishes with a detailed cost estimate example.

Chapter 5 is the last cost estimate chapter, focusing on the probabilistic cost estimate. It shows how to generate a probabilistic approach considering uncertainties

DOI: 10.1201/9781003402725-1

and risks. Also, it shows the definition and differences between contingency and management reserve.

The next block is the economic feasibility study covered by Chapters 6 and 7. It starts by showing each economic indicator used in the economic analysis, such as return on investment (ROI), payback period, return on capital employed (ROCE), net present value (NPV), profitability index (PI), and the break-even analysis. The seventh chapter exemplifies how to do an economic feasibility analysis and includes new concepts such as cash flow, inflation, and escalation. Also, it finishes by showing a probabilistic approach and a mitigated scenario for risks.

After the cost is estimated and the economic study is done, the cost should be controlled. This is discussed in Chapter 8. Several methods are discussed, such as planned versus performed, earned value management, indicators, S curve, and rundown curve.

Chapter 9 shows several topics that can be used through the lifecycle of assets, or during the cost estimate, economic feasibility study, and cost control. For this reason, they are covered separately. They are inflation, escalation, deprecation, benchmarking, sensitivity analysis, value engineering (it is most helpful during the planning phase), and readjustment.

Chapter 10 shows a correlation between sustainability and cost focusing in the carbon embodied estimation.

Although exercises and examples are based on real cases I worked on in the past, the company names and numbers used in the book are hypothetical. Also, simplifications are adopted for educational purposes. For example, a live detailed cost estimate can have thousands of materials and equipment to be costed, and this detail is beyond the purpose of the book.

An open tool could be used with the book as an option. The tool shows the resolution of the book's examples and exercises. The benefits are checking the formulas, analyzing different scenarios, changing the inputs, and verifying the step-by-step answer. Because it is an open tool that is not blocked, it becomes crucial to follow the tool instructions and create a backup file.

2 Basic Concepts

The chapter shows basic concepts, as listed below, and are used throughout the book. Also, they are the foundation of cost engineering.

- Cost × price
- Labor
- Material
- Equipment
- Direct cost and indirect cost
- Overhead
- Fixed, semi, and variable costs
- Scope and Work Breakdown Structure
- Cost Breakdown Structure
- CAPEX
- OPEX

2.1 COST × PRICE

Cost and price are different concepts. Cost is the money to produce a good or execute a service. Price is the amount of money asked for a product or service.

Figure 2.1 illustrates the cost concept. All resources used in the planning, design, construction, and commissioning phases are cost elements and compound the total plant cost. For example, the cost of a process plant is labor, material, equipment, construction equipment, other resources, and contingency. Labor, material, and equipment cost elements are detailed in the following sections. Also, additional costs such as the interest rate and overhead could be added according to the company's procedure

Price, as illustrated in Figure 2.2. 32, is the amount of money that the owner pays to build the facility including the profit and taxes.

2.2 LABOR

Labor costs include welders, plumbers, engineers, and all direct and indirect professional costs related to the service or good manufactory. Overall, the labor cost is estimated by Equation 2.1:

$$LC = Salary \times PQ \times (1 + FB) \times (1 + OT) \times Duration \quad \text{(Equation 2.1)}$$

Where:
LC = Labor cost
PQ = Professional's quantity

DOI: 10.1201/9781003402725-2

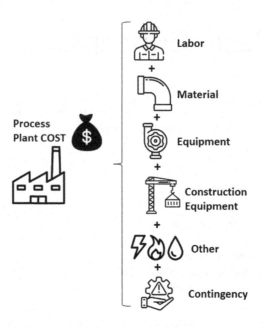

FIGURE 2.1 Cost of a process plant, icons made by Freepik and Smahicons from www.flaticon.com.

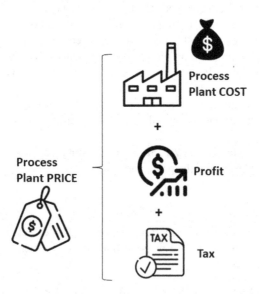

FIGURE 2.2 Price of a process plant, icons made by Freepik and Smahicons from www.flaticon.com.

FB = Fringe benefits
OT = Overtime

Fringe benefits are pension, medical and life insurance, and government benefits. They are different in each country and should be verified by current legislation. Also, it is measured by percentage.

Overtime is related to extra work. For example, a professional who works 40 hours per week or 8 hours per day will receive their hourly wage plus additional overtime if he works on Saturday. Overall, it is a factor of 50% or 100% applied in the extra hours but should be checked by legislation. Also, the fringe benefits could be considered in the extra hours, like in Equation 2.1. Each country's legislation should verify it because sometimes the benefits are not included in the overtime cost estimation. In this case, the formula for overtime cost should be:

$$LC = Salary \times PQ \times (1 + OT) \times Duration \qquad \text{(Equation 2.2)}$$

However, the book's examples consider it as included for simplification.

The salary could be estimated by the hour, week, month, or year. Salary references are obtained through company databases or specialized sites and are affected by qualification, location, market conditions, experience, company size, and position, as summarized in Figure 2.3.

FIGURE 2.3 Salary attributes and icons made by Freepik from www.flaticon.com.

A high experience means a better salary, and a degree suggests a wage improvement. Also, there is usually a region in each country where the wage is higher than others. For example, California overall has the highest salary in the US. The company size means that huge organizations tend to promote higher salary than small companies.

Market factor shows that by the Law of Demand and Supply, the wage tends to decrease if there are many professionals available and only a few jobs. Finally, professionals who report to the higher levels of the organization tend to have higher remunerations.

2.2.1 Labor Cost Estimate – Example 1

Company ABC needs assembly tubes, and the main points from the contract and regulation are listed below.

Contract Clauses

- Duration – 1.5 months or 6 weeks
- Bonus if the work finishes in 1 month = $ 25,000

Regulation

- Workweek – 40 hours
- Professionals can work 20 hours per week in overtime
- Fringe benefits = 50%
- Overtime factor = 1.5

Estimate the crew cost listed in Table 2.1 considering the duration of 6 weeks. Then, estimate if it should be profitable to execute the task in 1 month, considering the bonus and overtime.

Crew
The cost estimate should be calculated through Equation 2.1:

- Welder = $22 \times 3 \times (1 + 0.5) \times 40 \times 6 = $ US$ 23,760
- Plumber = $20 \times 4 \times (1 + 0.5) \times 40 \times 6 = $ US$ 28,800

TABLE 2.1
Labor Data – Example 1

Category	Hourly salary ($)	Quantity	Hours per week	Fringe benefits (%)	Duration (weeks)
Welder	22.00	3.00	40.00	50.0	6.00
Plumber	20.00	4.00	40.00	50.0	6.00
General Helper	15.00	5.00	40.00	50.0	6.00
Foreman	25.00	1.00	40.00	50.0	6.00

- General Helper = 15 × 5 × (1 + 0.5) × 40 × 6 = US$ 27,000
- Foreman = 25 × 1 × (1 + 0.5) × 40 × 6 = US$ 9,000
- Total = 23,760 + 28,800 + 27,000 + 9,000 = US$ 88,560

Considering significant figures, the result is US$ 88.6 thousand.

The second scenario is 1 month of duration and 20 hours of overtime per week. It is summarized in Table 2.2.

The cost estimate should be calculated into two steps through Equation 2.1:

Cost Estimate – First is calculated the work without overtime, as below.

- Welder = 22 × 3 × (1 + 0.5) × 40 × 4 = US$ 15,840
- Plumber = 20 × 4 × (1 + 0.5) × 40 × 4 = US$ 19,200
- General Helper = 15 × 5 × (1 + 0.5) × 40 × 4 = US$ 18,000
- Foreman = 25 × 1 × (1 + 0.5) × 40 × 4 = US$ 6,000
- Total = 15,840 + 19,200 + 18,000 + 6,000 = US$ 59,040, or US$ 59 thousand.

Second, the overtime is estimated considering that each professional works 20 hours and the overtime rate is 50%, as below:

- Welder = 22 × 3 × (1 + 0.5) × (1 + 0.5) × 20 × 4 = US$ 11,880
- Plumber = 20 × 4 × (1 + 0.5) × (1 + 0.5) × 20 × 4 = US$ 11,200
- General helper = 15 × 5 × (1 + 0.5) × (1 + 0.5) × 20 × 4 = US$ 13,500
- Foreman = 25 × 1 × (1 + 0.5) × (1 + 0.5) × 20 × 4 = US$ 4,500
- Total overtime = 11,880 + 11,200 + 13,500 + 4,500 = US$ 44,280 or US$ 44 thousand.

The total cost to this scenario is US$ 59,400 + US$ 44,280 = US$ 103,320 or US$ 100 thousand. However, a bonus of US$ 25,000 should be reduced from the total. Hence, the cost estimate is US$ 78,320 or US$ 78 thousand, considering significant figures, which is US$ 10 thousand lower than the option of 6 weeks without overtime.

One assumption is that the productivity index, discussed in Section 4.4.1, is the same in both scenarios, which could be an unrealistic hypothesis. Finally, a risk analysis could be done to ensure a complete study.

TABLE 2.2

Labor Data for a Task Duration of 4 Weeks – Example 1

Category	Hourly salary ($)	Quantity	Hours Per week	Fringe benefits (%)	Duration (weeks)	Overtime	Hours per week (overtime)
Welder	22.00	3.00	40.00	50	4.00	0.50	20.00
Plumber	20.00	4.00	40.00	50.0	4.00	0.50	20.00
General Helper	15.00	5.00	40.00	50.0	4.00	0.50	20.00
Foreman	25.00	1.00	40.00	50.0	4.00	0.50	20.00

2.2.1   FRINGE BENEFITS – EXAMPLE 2

Company ABC is studying the most profitable place to design their new asset between Green and Blue Cities/Teams. Define the Cost Estimate of each city/team and the best option from a cost perspective considering the data from Tables 2.3 and 2.4.

Design Team – Green City
Design Team – Blue City

The differences between the cities are the fringe benefits and the quantities.

Green City is proposing 32 professionals and Blue City 40. Also, the Blue City Benefits are half of Green City.

TABLE 2.3
Design Team – Green City – Example 2

Category	Monthly base wage ($)	Quantity	Fringe benefit (%)	Deadline (Months)
Manager	10,000	2.00	60.0	6.00
Electrical Engineer	7,000	5.00	60.0	6.00
Electronical Engineer	7,100	2.00	60.0	6.00
Mechanical Engineer	8,000	8.00	60.0	6.00
Naval Engineer	8,100	8.00	60.0	6.00
Comissioning Engineer	7,900	3.00	60.0	6.00
Planner	6,800	2.00	60.0	6.00
Project Manager	7,900	1.00	60.0	6.00
Project Control	6,800	1.00	60.0	6.00

TABLE 2.4
Design Team – Blue City – Example 2

Category	Monthly base wage ($)	Quantity	Fringe benefit (%)	Deadline (Months)
Manager	10,000	2.00	30.0	6.00
Electrical Engineer	7,000	7.00	30.0	6.00
Electronical Engineer	7,100	3.00	30.0	6.00
Mechanical Engineer	8,000	10.0	30.0	6.00
Naval Engineer	8,100	9.00	30.0	6.00
Comissioning Engineer	7,900	4.00	30.0	6.00
Planner	6,800	2.00	30.0	6.00
Project Manager	7,900	1.00	30.0	6.00
Project Control	6,800	2.00	30.0	6.00

The result, as shown in Table 2.5, is obtained through Equation 2.1 and applied for each professional. Blue City has a lower cost than Green City despite the bigger team. It occurs because the fringe benefits are considerably lower than Green City, and the question is if they are enough to pay all work regulations.

2.2.2 LABOR COST ESTIMATE – EXERCISE 1

Company ABC plans to repair a pipeline, and the duration is estimated at 8 weeks. Also, the maximum budget is $ 245,000. Calculate the cost estimate according to the data listed in Table 2.6. Is the budget enough?

Also, the company wants to mitigate a turnover risk by increasing the hourly salary by 10% and providing an additional benefit of 5%. After this extra cost, is the budget respected?

TABLE 2.5
Results Per City – Example 2

RESULTS	Cost Estimate Green City	Cost Estimate Blue City	Unit
Manager	192,000	156,000	US$
Electrical Engineer	336,000	382,200	US$
Electronical Engineer	136,320	166,140	US$
Mechanical Engineer	614,400	624,000	US$
Naval Engineer	622,080	568,620	US$
Comissioning Engineer	227,520	246,480	US$
Planner	130,560	106,080	US$
Project Manager	75,840	61,620	US$
Project Control	65,280	106,080	US$
TOTAL*	2,334,720 or 2,330,000	2,311,140 or 2,310,000	US$

*Second Result in each city uses three significant figures.

Crew

TABLE 2.6
Crew – Exercise 1

Category	Hourly base wage ($)	Quantity	Hours Per week	Fringe benefit (%)	Deadline (weeks)
Welder	20.00	6.00	40.0	50.0	8.00
Plumber	19.00	8.00	40.0	50.0	8.00
General Helper	14.00	10.00	40.0	50.0	8.00
Foreman	24.00	2.00	40.0	50.0	8.00

2.3 MATERIAL

According to Hastak, 2015, material can be divided into three categories:

- Raw materials
- Bulk materials
- Fabricated materials

Also, there are the engineered materials, but they are discussed in the equipment section.

Raw materials do not require a production process (e.g., coal, iron, sand, and gravel). They are the inputs to generate processed materials.

Bulk materials are related to availability. Overall, these materials are bought and delivered in a short time. For example, steel, nuts, bolts, gutter, and pipe are all examples of bulk materials.

As the name suggests, fabricated materials are bulk materials converted for a specific purpose. Figure 2.4 shows an example of pipes fabricated into spools and flanges at a pipe shop.

Material procurement should be done considering a group of factors:

- Contract duration
- Historical information
- Logistics
- Negotiation
- Quantity
- Storage
- Strategy
- Suppliers

The procurement strategy should be set at the beginning phase of a new investment. It defines whether one supplier will be responsible for all products, whether a pre-qualified supplier list should be established, or whether new suppliers should be sought.

The contract duration or delivery time has a price correlation because urgent demand tends to increase the price. If the material is negotiated, bought, and delivered with float time, the probability of reducing the price increases.

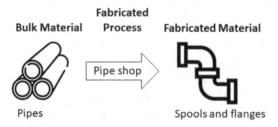

FIGURE 2.4 Bulk and fabricated material and icons made by Freepick from www.flaticon.com.

In addition, the quantities have a direct correlation with the price. Usually, the unit price tends to be higher for a small quantity rather than a higher one because of the economy of scale.

The logistics costs should be verified, but it is dicussed in Section 2.4.1.

Also, historical information helps to create cost references, shows lessons learned, and could be helpful to define the best strategy.

The taxes should be studied because the material could have a subsidy from a specific supplier/country. Hence, all quotations should inform the tax and any special tax conditions.

Negotiation should be done using the best practices, and previous planning should establish the goals and methods. Furthermore, the team must have expertise and skills to ensure that all requirements are attended to, and the process should be done following compliance and ethics guidance.

Potential suppliers should be periodically researched to increase the local offer and, if possible, international requests. The supplier database should be updated, and it must inform parameters such as quality, previous penalties, environmental, etc. Finally, suppliers should meet regulations, safety standards, relevant codes, have licenses to do business, and all necessary permits.

Storage is an attention point because a trade-off between just in time or storage for some period should be done.

One common and famous strategy is to obtain three quotations (minimum), called a price survey, from different suppliers that meet the requirements. The average or the lowest price can be used as a cost estimate, and the process should be recorded and explained/justified for future audits.

2.3.1 MATERIAL – EXAMPLE 3

Company Blue Sea is looking for materials, and three suppliers provide proposals in Table 2.7.

- What is the total price per supplier?
- What is the price per material through the average method (three prices)?
- What is the total price through the lowest method per material?

TABLE 2.7
Material and Unit Prices Per Supplier – Example 3

Material	Quantity	Supplier 1	Supplier 2	Supplier 3
Cable	50,000.0 m	US$ 5.00 / m	€ 4.00 / m	US$ 4.60 / m
Steel	20.0 ton	US$ 2000.00 / ton	€ 2100 / ton	US$ 2.50 / kg
Steel nuts	15,000.0 unit	US$ 1.00 / unit	€ 0.500 / unit	US$ 1.10 / unit
Site distance	-	500.00 km	300.00 miles	450.00 km
Delivery price (logistics)	-	US$ 5.00 /km	€ 3.50 / km	US$ 4.80 /km

Assumptions:

- It has not provided detailed information about each material for simplification.
- US$ 1.30 = € 1.00
- Delivery conditions (Incoterms) are the same for all suppliers.
- Tax is included in the price.
- 1 mile = 1.61 km
- 1 ton = 1,000 kg

First, calculate the price per supplier. Each material should be estimated by multiplying the rate per quantity and looking for adjustments as units and currency. Finally, the delivery price should be added.

Supplier 1 = 50,000 × 5 + 20 × 2,000 + 15,000 × 1 + 500 × 5 = US$ 307,500 or 308 thousand.

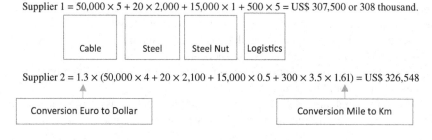

Supplier 2 = 1.3 × (50,000 × 4 + 20 × 2,100 + 15,000 × 0.5 + 300 × 3.5 × 1.61) = US$ 326,548

Supplier 2 = or US$ 327 thousand.
Supplier 3 = 50,000 × 4.6 + 20 × 2.5 × 1,000 + 15,000 × 1.1 + 450 × 4.8 = US$ 298,600 or 299 thousand.

Conversion kg to ton

Hence, the cheapest supplier to provide these materials is supplier 3. The second calculation is the price per material through the **average** method. Because the delivery price is not provided per material, it is not shown below:

Cable price = [50,000 × (5 + 4 × 1.3 + 4.6)]/3 = US$ 246,667 or US$ 247 thousand.
Steel price = [20 × (2,000 + 2,100 × 1.3 + 2.5 × 1,000)]/3 = US$ 48,200 or US$ 48.2 thousand.
Steel nut price = [15,000 × (1 + 0.5 × 1.3 + 1.1)]/3 = US$ 13,000 or US$ 13.0 thousand.

The last question is the lowest quotation per material. Table 2.8 summarizes the results per supplier and material. Cable by supplier 3 provides the lowest price, US$ 230,000. Supplier 1 has the lowest price for steel, and supplier 2 is the best option for steel nuts.

TABLE 2.8
Price Per Supplier and Material

	Supplier 1	Supplier 2	Supplier 3
Cable (US$)	250,000	260,000	230,000
Steel (US$)	40,000	54,600	50,000
Steel nut (US$)	15,000	9,750	16,500

TABLE 2.9
Data Exercise 2

	Unit Price	Observation
Supplier 1	US$ 900 per instrument	Delivery is included, and the local government subsidies 5.0% of the total cost.
Supplier 2	US$ 800 per instrument	It is an international supplier, and the delivery cost is US$ 10 per instrument. Also, there is an additional tax of 10.0%.
Supplier 3	US$ 890 per instrument	Delivery is US$ 1000, and the local government subsidies 5.0% of the total cost.

Hence, the total cost is US$ 279,750 or US$ 280 thousand. It is US$ 18.9 thousand lower than all materials provided by supplier 3. However, it is impossible to confirm it because the delivery price is not provided by material but only by the supplier.

2.3.2 MATERIAL – EXERCISE 2

The company Smart Box is looking for 3,000 instruments for their new manufacturing plant. There are three quotations, as shown in Table 2.9. What is the lowest price between them?

2.4 EQUIPMENT

Equipment is called engineered materials or designed materials, and vessels, pumps, heat exchangers, compressors, transformers, and towers are examples.

They require a complex manufacturing and design process, requiring many hours, and can be specified according to each project requirement.

A group of factors (Figure 2.5) should be considered when the cost estimate is done for equipment. The cost estimator must know them before starting the process. It is crucial to read requirements and specifications to understand what conditions are being proposed.

Manufacturing conditions are related to technical specifications, manufacturing deadlines, quantity, suppliers, and demand. If the complexity of equipment increases, the price tends to rise, too. For example, a water pump has a lower complexity and price than a turbine, as shown in Figure 2.6. Furthermore, the same equipment

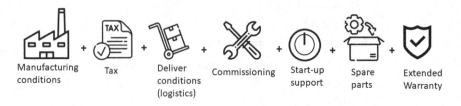

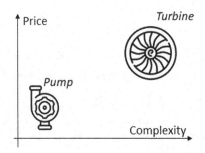

FIGURE 2.5 Factors that affect equipment cost and icons made by Freepick from www.flaticon.com.

FIGURE 2.6 Price × Complexity and icons made by Freepick from www.flaticon.com

using a specific material (e.g., carbon steel vs super duplex) could increase the price significantly.

Also, if the requirement is urgent, it implies higher costs. The quantity is important because the negotiation process is different if it is done for one pump or one hundred pumps (economies of scale), for example.

As illustrated in Figure 2.3, the Law of Demand and Supply shows that if there is a high demand for the equipment and not enough suppliers or production, the price tends to rise.

Tax should be verified because the quotation may or may not include the tax percentage. Hence, all offers must have an equal base to promote a fair comparison. Also, suppliers can have different taxes or subsidies from another country, implying a final price increment or reduction.

Commissioning and start-up are support services to ensure that equipment operates according to manufacturing conditions. This amount of money can be specified or input into the price.

Spare parts and warranty should respect the specification, and it may add additional charges to cover extended warranty, for example.

There are several methods to estimate the equipment cost:

- Quick method, as capacity factor
- Parametric equation
- Price survey
- Bottom-up

The first two are detailed in Chapter 3, and the supplier usually does the bottom-up because they know the item components and costs related to the manufacturing process.

The most common method is the price survey described in the previous section.

2.4.1 LOGISTICS AND DELIVERY CONDITIONS

Logistics costs are affected by several factors, such as the mode type (air, sea, or land), volume, weight, distance, quantity (batch), and delivery conditions. Also, if there are few companies to provide the service, the cost tends to increase.

It could be estimated by adopting a percentage of the total cost or using online calculators available on the internet, or a price survey. The attention point is the volume, and weight should consider the packing.

Overall, the delivery conditions are ruled by INCOTERMS®, which are maintained by the International Chamber of Commerce (ICC). These rules provide specific guidance to individuals participating in the import and export of global trade. These rules are crucial for equipment/material cost estimates.

Figure 2.7 shows the Incoterms 2020. The dashed bars show seller's obligation and the solid line bars show the buyer's obligation. Also, the Risk boxes show when the risk is transferred from seller to buyer. There are eleven possibilities of trade terms illustrated in the image, as below:

- EXW – ex works
- FCS – free carrier
- FAS – free alongside ship. Applicable only for sea transport
- FOB – free on board. Applicable only for sea transport
- CFR – cost and freight. Applicable only for sea transport
- CIF – cost, insurance and freight. Applicable only for sea transport
- CPT – carriage paid to
- CIP – carriage and insurance paid to
- DAP – delivered at place
- DPU – delivered at place unloaded
- DDP – delivered duty paid

FAS, FOB, CFR, and CIF only apply to sea transport. Others are applicable for any mode.

There are eight possible limits between the seller and buyer obligation: industry, first carrier, alongside ship (original port), on board, on arrival, alongside ship (destination port), destination place, and buyer.

For example, FCA means the seller will be responsible for the first carrier. From this step, the buyer will be responsible for the rest (cost, risk, and insurance). On the other hand, DDP means that the seller will be responsible for the cost, risk, and insurance until the destination.

CIP and CPT are the same, except for the insurance. In the CIP trade term, the seller is responsible for the insurance and CPT does not include goods insurance. In the same way, the primary difference between CFR and CIF is cargo insurance.

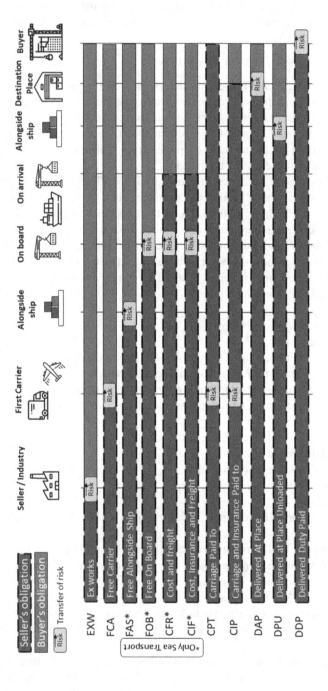

FIGURE 2.7 INCOTERMS® 2020. Icons made by Freepick from www.flaticon.com.

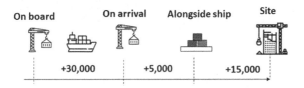

FIGURE 2.8   Additional delivery costs and icons made by Freepick from www.flaticon.com.

2.4.2   DELIVERY CONDITIONS – EXAMPLE 4

Two suppliers quoted a vessel. The specification informs that the supplier should provide spare parts for 1 year, extended warranty for 2 years, service for 1 month to support commissioning and start-up phases, and minimum delivery conditions is FOB. What is the lowest price?

Supplier A quotation: $ 1 million. DDP. 1 month of service/support to commissioning and start-up included. Extended warranty $ 100 thousand per year. Spare parts are included for 1 year. Taxes are included.

Supplier B quotation: $ 0.9 million. FOB. $ 5000 per day for service/support to commissioning and start-up. Extended warranty $ 100 thousand per year. Spare parts are included for 1 year. Taxes are included.

Assumption: if the delivery is not on-site, additional costs like the Figure 2.8 should be adopted to cover insurance, risk, and delivery costs.

Supplier A total cost is $ 1 million plus $ 200 thousand of warranty. And supplier B has a lower price, but the delivery condition is Free on Board (FOB) and commissioning and start-up costs are not considered. Hence additional costs should be added.

Additional delivery conditions: 30,000 + 5,000 + 15,000 = $ 50,000
Commissioning and start-up costs: 5,000 × 30 days = 150,000

Hence, the total cost of supplier B is:

Total cost supplier B: 900,000 + 50,000 + 150,000 + 200,000 = $ 1.3 million.
    So supplier A has the lowest price.

> Warranty costs

2.4.3   EQUIPMENT COST ESTIMATE – EXERCISE 3

Company Assembly Plus is estimating service to assemble one vessel and two pumps, as shown in Figure 2.9. Following the requirements below and using the three quotations as shown in Table 2.10, define the equipment cost estimate and the strategy. Should we adopt one supplier per equipment or one for all?

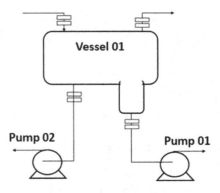

FIGURE 2.9   Scope diagram: Vessel 1 and Pumps 1 and 2 – Exercise 3

Scope and Requirements:

- Vessel 01
  Commissioning and start-up: 15 days,
  Spare parts: 1 year,
  Warranty: 5 years,
  Delivery conditions: Minimum FOB,
  Deadline: 1.5 years on the site,
  Special requirement: 3 inspections along with manufacturing,
- Pumps 01 and 02
  Commissioning and start-up: 1 month
  Spare parts: 2 years
  Warranty: 2 years
  Delivery conditions: Minimum FOB
  Deadline: 1 year on the site
  Special requirement: Additional sensors to monitor vibration, temperature,
      and pressure.

Quotations:

Assumption: if the delivery is not on-site, additional costs per equipment, like the
Figure 2.10, should be adopted to cover insurance, risk, and delivery costs.

2.5   DIRECT AND INDIRECT COST

The costs can be divided into direct and indirect. AACE International, 2022, defines
direct cost[1] as:

In construction, the cost of installed equipment, material, labor and supervision
directly or immediately involved in the physical construction of the permanent facility.
In manufacturing, service, and other non-construction industries: the portion of operating costs that is readily assignable to a specific product or process area.

TABLE 2.10

Quotations from Three Suppliers – Exercise 3

Supplier	Price	Offer description
Plus Equipment	Vessel 1 → US$ 1,500,000 Pump 1 → US$ 330,000 Pump 2 → US$ 450,000	Vessel 1→ US$ 5,000 per day for service/support to commissioning and start-up. Extended warranty $ 50,000 per year. Spare parts are included for 1 year. Duration manufacture estimation: 2 years, but the deadline could be attended with US$ 0.5 million additional costs. Hence, the price is $ 1.5 million considering the half million for acceleration. — no costs for inspections. FOB. Pumps 1 and 2 → US$ 5,000 per day for service/support to commissioning and start-up. Extended warranty $ 30,000 per year. Spare parts are included for 2 years. Duration manufacture estimation: 1 year. US$ 50,000 for sensors per pump. FOB.
Power and Material	Vessel 1 → US$ 2,200,000 Pump 1 → US$ 500,000 Pump 2 → US$ 600,000	Vessel 1→ service/support to commissioning and start-up are included. Extended warranty $ 50,000 per year. Spare parts are included for 1 year. Duration manufacture estimation: 1.5 years — no costs for inspections. DDP. Pumps 1 and 2 → service/support to commissioning and start-up are included. 2 years of extended warranty. Spare parts are included for 2 years. Duration manufacture estimation: 1 year. US$ 10,000 for sensors per pump. DDP.
Solutions	Vessel 1 → US$ 1,800,000 Pump 1 → US$ 430,000 Pump 2 → US$ 550,000	Vessel 1→ service/support to commissioning and start-up are included. Extended warranty $ 25,000 per year. Spare parts are included for 1 year. Duration manufacture estimation: 1.5 years — no costs for inspections. FOB. Pumps 1 and 2 → service/support to commissioning and start-up are included. 2 years of extended warranty. US$ 10,000 spare parts per year. Duration manufacture estimation: 1 year. Sensors per pump are included. FOB.

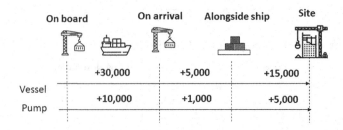

FIGURE 2.10 Additional costs per equipment – Exercise 3 and icons made by Freepick from www.flaticon.com

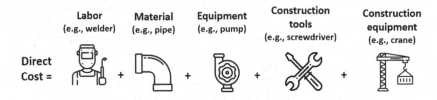

FIGURE 2.11 Direct cost and icons made by Freepick from www.flaticon.com

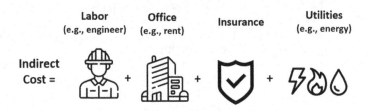

FIGURE 2.12 Indirect cost and icons made by Freepick from www.flaticon.com

And indirect costs[2] is

> in construction, (field) indirects are costs which do not become a final part of the installation, but which are required for the orderly completion of the installation and may include, but are not limited to, field administration, direct supervision, capital tools, startup costs, contractor's fees, insurance, taxes, etc. In manufacturing, costs not directly assignable to the end product or process, such as overhead and general purpose labor, or costs of outside operations, such as transportation and distribution. Indirect manufacturing cost sometimes includes insurance, property taxes, maintenance, depreciation, packaging, warehousing and loading.

The definition could be different according to industry type. Figure 2.11 shows an example of the direct cost of building a factory. All resources are directly used for its construction and completion.

Indirect cost is not directly associated with an activity. For example, Figure 2.12 shows an example of the indirect cost of building a factory. Hence, indirect labor costs, such as engineers, managers, office costs, rent, desks, IT costs, utilities, and insurance costs are examples of indirect costs.

2.6   OVERHEAD

Overhead is related to tasks not identified as part of the work but intrinsic. For example, the Human Resources (HR) department must hire, pay, and provide all support to the team to execute the task properly. Also, HR supports all contracts and works developed by construction, engineering, or manufacturing companies.

Other examples of overhead are the legal, finance and accounting, IT, estimating, and sales departments.

The overhead cost could be slipped into the portfolio by applying a percentage. After the direct and indirect cost is estimated, a rate is added. This percentage can be fixed or variable according to factors such as complexity. These two methods are discussed in the following section.

2.6.1   FIXED AND VARIABLE PERCENTAGE – EXAMPLE 5

Company Blue estimates that overhead costs are $ 170,000 in the current year. Also, they have six projects in the period, as listed below.

Direct and Indirect Cost Estimate:

Project 1 = $ 3,000,000
Project 2 = $ 2,000,000
Project 3 = $ 1,000,000
Project 4 = $ 700,000
Project 5 = $ 500,000
Project 6 = $ 400,000

The value estimated per project is the direct and indirect cost to execute the contract scope. However, how does the company split the overhead through the projects?

Two scenarios are analyzed.

First, the company used the fixed percentage method to deal with this cost. A percentage is applied to the cost estimate to determine the overhead cost.

Project 1 = $ 3,000,000 × 2% = $ 60,000
Project 2 = $ 2,000,000 × 2% = $ 40,000
Project 3 = $ 1,000,000 × 2% = $ 20,000
Project 4 = $ 700,000 × 2% = $ 14,000
Project 5 = $ 500,000 × 2% = $ 10,000
Project 6 = $ 400,000 × 2% = $ 8,000

Hence, the total is insufficient ($ 152 thousand < $ 170 thousand).

The second scenario is a variable percentage. It can be a high demand for specific projects (e.g., a high level of procurement or legal demand). In the example, Projects 1, 2, and 3 estimate a high overhead demand, so a different percentage (3%) should be used. Also, other projects expect a low overhead demand, and a low rate of 0.5%, is adopted. The result shows that costs are covered using this method.

Project 1 = $ 3,000,000 × 3% = $ 90,000
Project 2 = $ 2,000,000 × 3% = $ 60,000
Project 3 = $ 1,000,000 × 3% = $ 30,000
Project 4 = $ 700,000 × 0.5% = $ 3,500
Project 5 = $ 500,000 × 0.5% = $ 2,500
Project 6 = $ 400,000 × 0.5% = $ 2,000

Hence, the total is enough ($ 188 thousand > $ 170 thousand).

The method selection should be made considering the portfolio and industry specifications.

2.7 FIXED, SEMI AND VARIABLE COSTS

The cost can be classified as fixed, semi-variable, and variable, which is helpful for operation or production analysis.

Fixed costs do not relate to the amount of work or production done. For example, tools such as hammers and welder machines have the exact cost if used 5 or 10 hours per day. Also, the office or factory rent is equal if operating 8 or 24 hours daily. Hence, these costs do not directly relate to the work or production and are classified as fixed costs.

Variable costs mean the cost increases if the amount of work or production increases – e.g., direct labor cost growth according to the number of hours increases. If the welder increases the hours per day from 8 to 10 hours, the wage rises too. Also, raw material grows directly with production increment.

Semi-variable costs have a fixed and a variable component, so it is a hybrid classification. The energy cost for a factory could be a semi-variable cost example. If there is no production, there is a minimum energy cost to maintain the safety and security conditions. In addition, if the output rises, the energy cost increases.

Figure 2.13 shows the fixed, variable, and semi-variable graphs. Fixed costs are a parallel line with production, it does not change with production raising. The variable costs have an angle, α, which is related to the production and cost increase. Finally, the semi-variable has both components of fixed and variable.

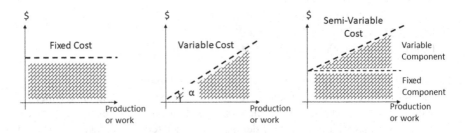

FIGURE 2.13   Fixed, variable, and semi-variable cost graphs

2.7.1 CLASSIFICATION ANALYSIS – EXAMPLE 6

A factory where the production (P) is 250 tons/month of a specific product has the cost segregation as Table 2.11.

- What is the total operating cost per month?
- What is the minimum cost (production = zero)?
- Which has the highest cost between labor, material, equipment, and other categories?
- The production manager aims to reduce the total operation cost to US$ 280 thousand per month. He is planning a 10.0% of reduction in the material cost. Is the plan satisfactory?

The total operating cost is estimated by adding the fixed, variable, and semi-variable cost, like below:

- Fixed cost (FC) = 10,000 + 40,000 + 25,000 + 6,000 = $ 81,000
- Variable cost considering P = 250 ton/month

$$VC = [(100 \times 250) + (500 \times 250) + (5 \times 250)] = \$ 151,250$$

- Semi-variable cost (SVC) considering P = 250 ton/month

$$SVC = [(50 \times 250 + 30,000) + (20 \times 250 + 25,000) + (2 \times 250 + 2,000)] = \$ 75,000$$

- Total operation cost = 81,000 + 151,250 + 75,000 = US$ 307,250 or US$ 307 thousand.

TABLE 2.11
Operation Cost – Example 6

Classification	Category	Value or Equation
Fixed Cost (FC)	Labor	$ 10,000
	Material	$ 40,000
	Equipment	$ 25,000
	Other	$ 6,000
Variable Cost (VC)	Labor	$100 \times P$
	Material	$500 \times P$
	Other	$5.00 \times P$
Semi-Variable Cost (SVC)	Material	$50.0 \times P + 30,000$
	Equipment	$20.0 \times P + 25,000$
	Other	$2.00 \times P + 2,000$

The minimum cost occurs when the production is zero. Hence, the variation cost is zero, the fixed cost is not modified, and the semi-variable is calculated as below:

- Semi-variable cost (SVC) considering $P = 0$ ton/month

$$SVC = [(50 \times 0 + 30,000) + (20 \times 0 + 25,000) + (2 \times 0 + 2,000)] = \$ 57,000$$

- Total $= 81,000 + 0 + 57,000 = \$ 138,000$ or US$ 138 thousand.

Total cost by category:

- Material $= FC + VC + SVC = 40,000 + 500 \times 250 + 50 \times 250 + 30,000 = \$ 207,500$
- Labor $= FC + VC = 10,000 + 100 \times 250 = \$ 35,000$
- Equipment $= FC + VC = 10,000 + 100 \times 250 = \$ 35,000$
- Other $= FC + VC + SVC = 6,000 + 5 \times 250 + 2 \times 250 + 2,000 = \$ 9,750$

Hence, the material category has the highest cost.

The last question is a plan to reduce 10% of the material cost to achieve a new total operation cost of $ 280,000.

$$\text{Material reduction} = \$ 207,500 \times (1 - 0.1) = \$ 186,750$$

Hence the new total includes labor, material, equipment, and other categories.

$$\text{Total operation cost} = 35,000 + 186,750 + 35,000 + 9,750 = \$ 286,500$$

Unfortunately, the plan is not achieving the reduction goal.

2.8   SCOPE AND WORK BREAKDOWN STRUCTURE

Scope is all work that should be done in a project and should be documented. The critical point is to ensure that the scope is identified and communicated to all stakeholders. Also, the exclusions should be highlighted to avoid misunderstanding. One definition of the products of scope is the Work Breakdown Structure.

The Work Breakdown Structure (WBS)is a framework to organize and give a hierarchy to a project. The WBS should be divided from the asset level (final product), top level, to lower levels to show each deliverable or work pack. It defines, in particular, the work necessary to accomplish a program/project's objectives. In addition, it is a communication tool because it provides a clear picture of what needs to be completed and the work that will be done.

WBS has several functions:

- Segmenting the project into identifiable and manageable units
- Identifying contracted, projected, and actual costs and the associated schedule components of the entire project

- Integrating cost and schedule for planning and controlling project progress and
- Permitting summarization of cost and schedule status for management reporting

Also, the WBS isnot a schedule, an organization chart, and a list of tasks. Figure 2.14 shows the WBS example. The first level shows the final objective, which is a new industrial plant. The second level splits the asset into major areas: onsite and offsite. The third level divides the areas into deliverables, and the last level provides the work packs. The WBS can have many levels, but it should allow integration with the schedule and control system because many levels increase complexity.

Each level of the WBS should respect the 100% rule. It means that all percentages from the following level (child) totalizes 100%. For example:

2° Level – Offsite (parent) = 3° Level – Roads, 25% + Parking, 25%, + Utilities, 50% (child)

It is recommended that the WBS has a dictionary to describe in brief narrative format what work is to be performed in each WBS element.

WBS is a live document, meaning that the WBS and dictionary must be updated if a change occurs. Furthermore, if possible, the standardized WBS is necessary for an organization because it facilities the collection and sharing of information.

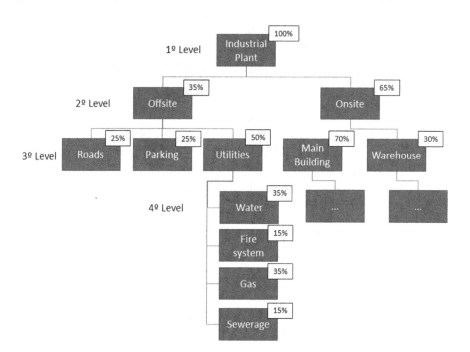

FIGURE 2.14   WBS example

In most cases, WBS, according to GAO, 2020, should be a product-orientated type because this allows a project manager to more precisely identify which components are causing the cost or schedule overruns and mitigate the root cause more effectively.

2.9   COST BREAKDOWN STRUCTURE

Cost Breakdown Structure (CBS) is defined as "A hierarchical structure that divides budgeted resources into elements of costs, typically labor, materials and other direct costs. The lowest level, when assigned responsibility, typically defines a cost centre" (AACE International, 2022).

Figure 2.15 provides an example where the cost is a breakdown of direct or indirect, then as labor, material, or equipment, and finally, the last level for labor's cost is detailed.

CBS provides cost analyses through each cost centre because the company can understand the behavior of each cost category. It means that it permits us to know if the cost increases or decreases along with the operation or implementation phases.

2.9.1   WBS AND CBS – EXAMPLE 7

Sky Blue Company wants to analyse the cost of the Process Plant X project. Through the integration of WBS and CBS, answer the following questions:

- What is the project cost estimated?
- What is the off-site cost estimated?

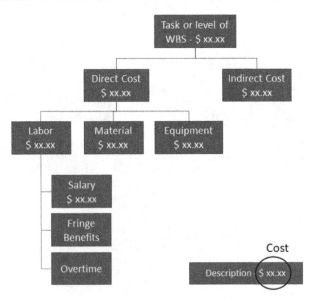

FIGURE 2.15   CBS example

- What is the storage tank cost estimated?
- What is the civil construction cost estimated for the storage tank?
- What is the direct cost estimated of civil construction for the storage tank?
- And what is the labor cost for the direct cost?

The answers are obtained through the analyses of the WBS and CBS integration. Figure 2.16 shows the result between the WBS and CBS,. Not all components are shown in the figure, only the necessary levels and elements to answer the example questions. Also, each branch follows the 100% rule, meaning each son's sum should totalize 100%. For example, the sons of Project X element, which estimate cost at $ 10 million, must totalize $ 10 million with on-site $ 6 million plus off-site 4 million.

- What is the project cost estimated? Level 1 = $ 10 million
- What is the off-site cost estimated? Level 1.2 = $ 4 million or 40% of total cost
- What is the storage tank cost estimated? Level 1.2.2 = $ 1 million or 25% of off site cost
- What is the civil construction cost estimated for the storage tank? Level 1.2.2.2 = $ 300 thousand or 30% of storage tank cost
- What is the direct cost estimated of civil construction for the storage tank?

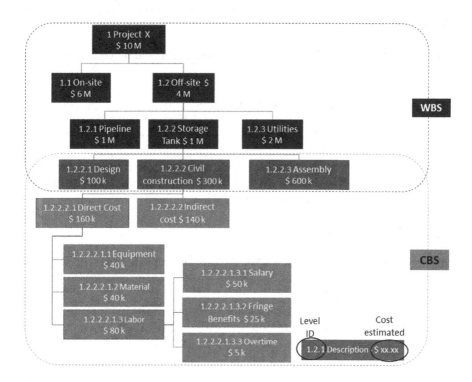

FIGURE 2.16   WBS and CBS integration Example 7

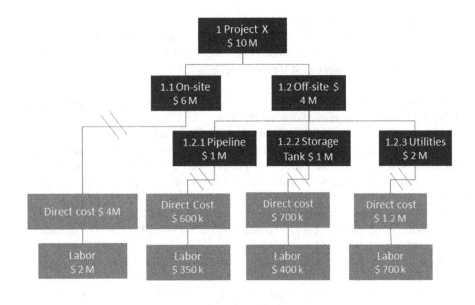

Not all levels/breakdowns are shown.

FIGURE 2.17   Direct labor cost – Project X – Example 7

Level 1.2.2.2.1 = $ 160 thousand or 53% of the civil construction cost for the storage Tank

- And what is the labor cost for the direct cost?

Level 1.2.2.2.1.3 = $ 80 thousand or 50% of the direct cost for civil construction

It is not shown in Figure 2.16, but it is possible to discover project direct labor costs. It occurs because all major areas have labor costs. Consequently, it can be summarized as $ 3.45 million, according to Figure 2.17.

2.9.2   WBS AND CBS – EXERCISE 4

Project Flat A's estimated cost is $ 100 thousand, and the WBS and CBS are illustrated in Figure 2.18. Through the figure, answer the following questions:

- What is the cost of the kitchen, bathroom, bedroom, and living room?
- What does the total civil construction cost?
- What does the kitchen labor cost?

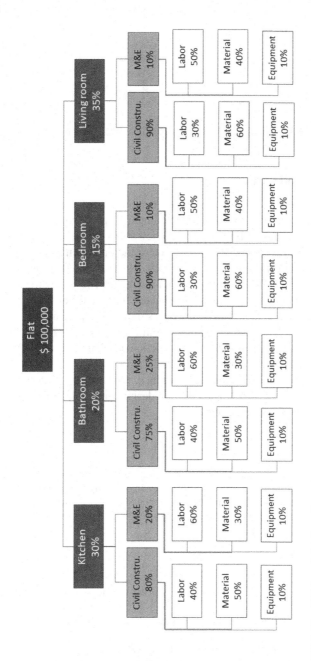

FIGURE 2.18 WBS and cbs flat – Exercise 4

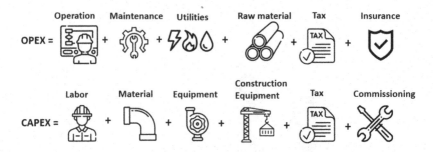

FIGURE 2.19 OPEX and CAPEX example and icons made by Freepick from www.flaticon.com.

2.10 CAPEX AND OPEX

Capital Expenditures (CAPEX) are related to investments to build or buy a new asset, facility, or equipment to improve the company's production.

Operating Expenses (OPEX) are the money spent on the operations activities, such as utility bills, raw materials, maintenance, and operation costs.

There are several differences between both concepts. OPEX is accounted for in the current year, but CAPEX is accounted for during the lifespan while assets depreciate. Also, the OPEX Tax treatment is deducted in the current tax year, while the CAPEX tax treatment is deducted over time as the asset depreciates.

Finally, Figure 2.19 shows an example from the oil and gas industry. It demonstrates that CAPEX is applicable for oil production development, and OPEX is for oil operation/extraction-related costs.

OPEX is the sum of operation, maintenance, utilities, raw material, taxes, and other costs. Otherwise, CAPEX is the sum of design, material, equipment, construction equipment, taxes, and other costs.

NOTES

1. Reprinted with permission from AACE International. Check the website for the latest versions (https://web.aacei.org/resources/cost-engineering-terminology).
2. Reprinted with permission from AACE International. Check the website for the latest versions (https://web.aacei.org/resources/cost-engineering-terminology).

BIBLIOGRAPHY

AACE International, 2022. *Recommended Practice 10S-90 Cost Engineering Terminology.* Available at: https://web.aacei.org/resources/cost-engineering-terminology. [Accessed: 8 January 2023].

GAO, 2020. Cost Estimating and Assessment Guide. Available at: https://www.gao.gov/products/gao-20-195g. [Accessed: 15 December 2022].

Hastak, M., 2015. *Skills & Knowledge of Cost Engineering*, 6th Edition. AACE International.

Incoterms®, 2020. Available at: https://iccwbo.org/resources-for-business/incoterms-rules/incoterms-2020/. [Accessed: 2 January 2023].

EXERCISE ANSWERS

1- Cost estimate $ 220,800. Budget is enough.
   Scenario with additionals: $ 250,976. Budget is not enough.
2- Lowest price, supplier 1, $ 2,565,000.
3- Solutions supplier has the lowest total price $ 3,027,000. Vessel lowest price
   $ 1,875,00 (plus equipment). Pump 1 and 2 have the lowest price $ 466,000
   and 586,000 (Solutions). The strategy using different suppliers is the most
   economical.
4- Kitchen $ 30,000, bathroom $ 20,000, bedroom $ 15,000, and living room
   $ 35,000. Total Civil Construction $ 84,000. Kitchen labor cost $ 13,200.

3 Cost Estimate – Quick Methods

Cost estimate development is a crucial step along with the project-planning phase. The cost estimate is important to evaluate the feasibility of a project, screen project alternatives, and evaluate the cost impacts of design alternatives.

There are three main cost estimation methods: quick or conceptual, detailed or bottom-up, and probabilistic. This chapter describes the quick methods, such as end-product units, physical dimensions, capacity factor, factor method, and parametric equation.

Like the name, these methods provide a fast result using a few inputs from the project. For example, it is possible to define the house cost by the area or square meter. For this reason, these methods are called analogy methods. In addition, they are applied in the initial phase when the project has a low maturity and high uncertainties, so quick methods are recommended for the conceptual planning phase.

Expert opinion is one quick method where experts define the cost estimate using their expertise. It is recommended that two or three opinions be collected, and an average of their opinion is adopted. However, this method has a low level of confidence.

The main disadvantage of the quick methods is the accuracy range because quick methods have a wider accuracy range than the detailed method, as shown in Figure 3.1.

3.1 PHYSICAL DIMENSIONS

Physical dimensions (PD) method is based on the area, volume, length, etc. For example, the cost per square meter can define the building cost.

Equation 3.1 shows the general formula for the PD method.

$$Cost = PD \times K \qquad \text{(Equation 3.1)}$$

Where:
PD = Quantities of physicals dimensions (e.g., m, m^2, or m^3)
K = Cost of one physical dimension

Historical information plays an important role because the analogy is made through past projects. For example, the historical information provides that the square meter cost is US\$ 1,000, so for one apartment in which the area is 100 m^2, the cost estimate is US\$ 100,000. However, the cost estimate is an order of magnitude because it does not consider several factors, such as quality, location, sunlight,

 DOI: 10.1201/9781003402725-3

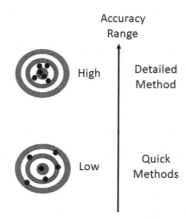

FIGURE 3.1 Accuracy range by method

facilities, noise, etc. Two apartments in the same area and city can have a considerable price variation, but the order of magnitude tends to remain the same.

3.1.1 OFFICE COST ESTIMATE – EXAMPLE 1

A business is looking for a new office of 90 m². The budget is US$ 630,000, and through the data below, what cities in Table 3.1 should be recommended?

Through Equation 3.1, the office cost can be estimated per city.

Beijing ➜ 10,000 × 90 = US$ 900,000
Canberra ➜ 3,200 × 90 = US$ 288,000
London ➜ 8,100 × 90 = US$ 729,000
Madrid ➜ 3,100 × 90 = US$ 279,000
New Delhi ➜ 1,400 × 90 = US$ 126,000
New York ➜ 8,000 × 90 = US$ 720,000
Paris ➜ 10,200 × 90 = US$ 918,000
Rio de Janeiro ➜ 1,500 × 90 = US$ 135,000
Rome ➜ 3,500 × 90 = US$ 315,000
Tokyo ➜ 6,800 × 90 = US$ 612,000

Then, the recommended cities should have a cost estimate below the budget of US$ 630,000. Figure 3.2 summarizes the result showing that Canberra, Madrid, New Delhi, Rio de Janeiro, Rome, and Tokyo are below the budget.

3.1.2 PIPELINE COST ESTIMATE – EXAMPLE 2

A new gas pipeline between Lisbon and Rome has been studied, and it has two options: one by the ocean/land using the latest technology, with a total distance of

TABLE 3.1

Price per Square Meter per City – Example 1

	City	Cost per square meter (US$/m²)
1	Beijing	10,000
2	Canberra	3,200
3	London	8,100
4	Madrid	3,100
5	New Delhi	1,400
6	New York	8,000
7	Paris	10,200
8	Rio de Janeiro	1,500
9	Rome	3,500
10	Tokyo	6,800

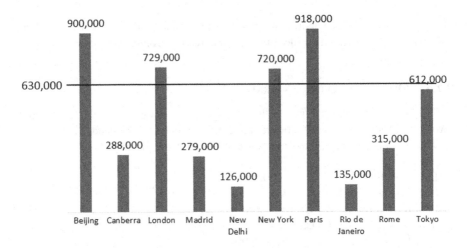

FIGURE 3.2   Cost estimate per city, and the black line represents the budget

1,865 km, and a second only by the ground through traditional technology, in which the length is 2,660 km. Figure 3.3 summarizes the two possible routes.

What is the best cost option considering the metrics in Table 3.2:

Through equation 3.1, the total cost per type is

Traditional technology and just land route ➜ 400,000, × 2,660 = US$ 1.064 billion

New technology and using land and ocean routes ➜ 520,000 × 1,865 = US$ 0.970 billion

Hence, the new technology is cheaper than the traditional, despite the cost per km being higher. Of course, it is a high-level cost estimate, and both a detailed study and a bottom-up estimate should be done to determine if the project will be executed.

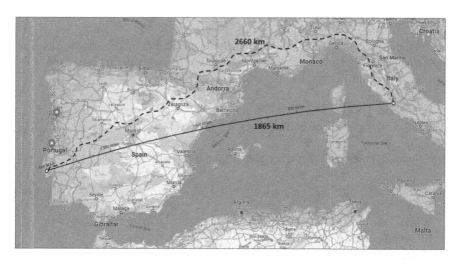

FIGURE 3.3 Two possible routes for a theoretical gas pipeline between Lisbon and Rome, map adapted by author from Google maps (download in 07/01/2022)

TABLE 3.2
Price per Km and Technology Type – Example 2

Type	Cost (US$)
Traditional technology and just land route	400,000 per km
New technology and using land and ocean routes	520,000 per km

TABLE 3.3
Capacity and Cost Metric – Exercise 1

	Capacity (m³/day)	Cost (US$ per m³/day)
Process plant A – Green City	2,000	1,000,000
Process plant B – Big City	1,000	800,000

3.1.3 Process Plant Cost Estimate – Exercise 1

Company ZK is in the conceptual phase of a new process plant. The capacity required is 2,000 m³ per day, and there are two options, as detailed in Table 3.3.

The maximum capacity of plant B is 1,000 m³/day, which means that any capacity over that amount requires an additional plant.

What is the cost estimate per plant type? What is the best option?

Observation: If two or more plants are installed in the Big City (plant B), an additional cost of US$ 100 million to cover interconnections should be included.

3.2 END-PRODUCT UNITS

The end-product Unit (EPU) method is similar to the PD method but is based on units completed. For example, a hotel cost of 100 rooms could be estimated if the cost of one room is known, like below:

Room cost: US$ 20,000
Hotel cost estimate: $100 \times 20,000 = $ US$ 2,000,000

Equation 3.2 shows the general formula for the EPU method.

$$\text{Cost} = \text{EP} \times \text{U} \qquad \text{(Equation 3.2)}$$

Where:
EP = Quantities of end-product
U = Cost of one unit

The method presumes that similar projects provide references and historical information. Also, one disadvantage is that the method does not consider economies of scale. For example, if the hotel proposed above has 500 rooms, the room cost should decrease. Furthermore, the method is not considering the time because if the reference is a long time away, it should be adjusted before being used.

3.2.1 RENEWABLES PLANT COST ESTIMATE – EXAMPLE 3

Two 10 MW renewable energy facilities have been studied in the conceptual phase, and the options are a solar farm or a wind farm, as illustrated in Figure 3.4. Determine the cost estimate of each option to support the decision process, considering the data in Table 3.4.

First, the number of solar panels and wind turbines should be determined:

Number solar panels = solar farm capacity/solar panel capacity
Number solar panels = 10,000,000/250 = 40,000 PV panels

FIGURE 3.4 Wind turbine (left) and PV panel (right), icons made by Freepick from www.flaticon.com

TABLE 3.4

Data Example 3

	Unit	Solar farm	Wind farm
Cost per PV panel or turbine	US$	210	10,000
Capacity per PV panel or turbine	kW	0.25	10
Farm capacity	kW	10,000	10,000

Number wind turbines = wind farm capacity/panel capacity
Number wind turbines = 10,000,000/10,000 = 1,000 turbines

Second, the cost estimated is calculated.
Solar farm cost = 40,000 × 210 = US$ 8,400,000
Wind farm cost = 1,000 × 10,000 = US$ 10,000,000

Hence, the solar farm is a lower result than the wind farm, but the decision process should consider several factors such as operation, maintenance, regulation, risks, etc. In addition, the cost estimate is not considering all costs as auxiliary equipment. It provides an order of magnitude.

3.2.2   BUILDING COST ESTIMATE – EXAMPLE 4

A construction company is analyzing two options for a new ten floor building. One option is three apartments per floor, of which the cost per unit is US$ 400,000, but the quality is standard. The second option is one apartment per floor which cost per unit $ 1 million, but it has a high-quality level. Which option is more expensive?

The cost estimate per option is calculated through Equation 3.2:

Option 1 = 10 floors × 3 apartments × US$ 400,000 = 12,000,000
Option 2 = 10 floors × 1 apartment × US$ 1,000,000 = US$ 10,000,000

Hence, the high-quality building shows a lower cost despite the apartment cost being 2.5 times higher than the standard quality apartment.

3.2.3   BUILDING COST ESTIMATE – EXERCISE 2

Using data from Example 1, Section 3.1.1, define the most expensive/cheapest city cost and the average for a building with four floors and four apartments per floor if each unit has 100 m².

3.3   CAPACITY FACTOR

The method is based on a comparison between similar plants but different capacities. Equation 3.3 shows the general formula for the capacity factor.

Cost new plant = Cost similar plant × (Cap_new / Cap_similar)^e (Equation 3.3)

Where:
Cap_similar = Capacity of the similar plant,
Cap_new = Capacity of new plant,
e = Exponent

Historical information is crucial to make the comparison and to generate the exponent. The exponent can differ per industry, but it oscillates between 0.5 and 0.85 (AACE International).

Also, adjustments of location and time should be considered when the method is applied because a similar plant built 10 years ago has a different cost now.

3.3.1 Process Plant Cost Estimate – Example 5

Define the cost for a process plant whose capacity is 100 tons per day through a similar plant of 50 tons per day which costs US$ 100 million. Both plants are located in the same city, and the reference was built 2 years ago. This means that escalation should not be considered in a low inflation scenario.
Also, e = 0.60
   Through Equation 3.3:

   Cost new plant = 100,000,000 × (100/50)$^{0.6}$ = US$ 151,571,657.
   Considering significant figures, the result is US$ 150 million.

One interesting point is that the cost is not doubled, although the new plant is double the reference. It occurs because of the economies of scale. For example, both plants could have one feedwater pump, but the new one has higher power or the operation is executed for more hours per day.

3.3.2 Location, Inflation, and Escalation – Example 6

Define the cost estimate using the last example, but the reference plant was built 5 years ago in a different city, called City A, referring to (Figure 3.5). The cost factor from City A to City B where the new plant will be constructed is 10%. Also, the escalation in the period is 30%.
   The first step is to update the reference cost using the time and location factors:

Original cost × (1 + escalation factor) x (1 + location factor)
Adjusted similar plant cost = 100,000,000 × (1 + 0.3) x (1 + 0.1) = US$ 143,000,000.

Through Equation 3.3:

   Cost new plant = 143,000,000 × (100/50)$^{0.6}$ = US$ 216,747,469.
   Considering significant figures, the result is US$ 217 million.

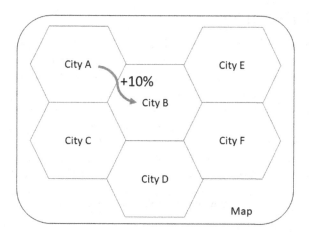

FIGURE 3.5 Map – Example 6

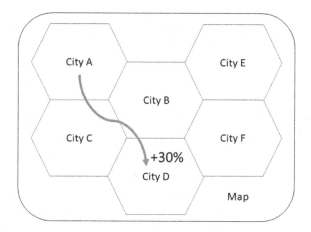

FIGURE 3.6 Map – Exercise 3

The cost estimate shows that the result is double than cost reference. Despite the economies of scale, escalation (time factors) and location cost considerably affect the final estimate.

3.3.3 Process Plant Cost Estimate – Exercise 3

Define the cost for a process plant whose capacity is 1,000 kg per hour through a similar plant of 3,000 kg per hour, which costs US$ 10 million. The reference plant was built 2 years ago in a different city, City A, as shown on the map (Figure 3.6). The cost factor from City A to City D, where the new plant will be constructed, is 30%. Also, the escalation in the period is 12%.

$$\text{Exponent } (e) = 0.5$$

3.4   FACTOR METHOD

The factor method is based that the total cost can be estimated from a cost element. For example, the total cost of a process plant is calculated from equipment cost. The factor should be defined through historical information, expertise, and mathematical analysis.

Furthermore, all factors should be estimated according to the industry and country that they are applied and must be updated periodically.

Equation 3.4 shows the factor method basic formula:

$$\text{Cost estimate} = \text{Primary cost} \times \text{factor} \qquad \text{(Equation 3.4)}$$

The factors used in the following examples and exercise are hypothetical and aim to illustrate the method.

3.4.1   Factor Cost Estimate – Example 7

Define the process plant cost considering that the factor adopted in this industry is 3.8, and the equipment total cost is estimated at US$ 50 million.

Through Equation 3.4:

$$\text{Cost estimate} = 50 \text{ million} \times 3.8 = \text{US\$ 190 million}$$

Hence, it shows that direct and indirect labor, material, and construction equipment have a cost of US$ 190 million.

3.4.2   Facility Cost Estimate – Example 8

Estimate the facility cost considering the factors per topic shown in Table 3.5:

Each area has a respective cost reference that defines the total cost. For example, the direct labor cost is the base to estimate the design, project management, and

---

TABLE 3.5

Factors per Subject, Cost Element and Their Respective Cost Estimate – Example 8

Description	Factor	Cost element description	Cost (US$)
Project management	1.2	Direct labor cost of project management	1,225,000
Design	1.4	Direct labor cost of design	1,750,000
Civil construction	4.7	Total cost of m³ concrete	4,100,000
Piping	3.4	Total pipe cost	7,000,000
Mechanical	3.7	Total equipment cost	9,800,000
Electrical and instrumentation	2.5	Total electrical equipment and instruments costs	4,600,000
Commissioning	2.1	Direct labor cost of commissioning	2,500,000

commissioning. It occurs because this rubric is a crucial input to calculate the topic cost.

Equation 3.4 should be applied for each topic:

- Project management = 1.2 × 1,225,00 = US$ 1,225,000
- Design = 1.4 × 1,750,000 = US$ 2,450,000
- Civil construction = 4.7 × 4,100,000 = US$ 19,270,000
- Piping = 3.4 × 7,000,000 = US$ 23,800,000
- Mechanical = 3.7 × 9,800,000 = US$ 36,260,000
- Electrical and instrumentation = 2.5 × 4,600,000 = US$ 11,500,000
- Commissioning = 2.1 × 2,500,000 = US$ 5,250,000

The facility cost is estimated by adding each term, like below:

$$\text{Facility cost} = 1,225,000 + 2,450,000 + 19,270,000 + 23,800,000 + 36,260,000 + 11,500,000 + 5,250,000$$

$$\text{Facility cost} = \text{US\$ } 100,000,000$$

3.4.3 PROCESS PLANT COST ESTIMATE – EXERCISE 4

Two countries have been analyzed to build a new plant but have different factors and equipment costs. The differences occur because of material and labor costs, fees, and subsidies. Using the factor method, define what country has the lowest cost considering the data in Table 3.6.

3.5 PARAMETRIC EQUATION

The parametric equation method is based on a mathematical model or equation representing a cost relationship between features such as the power and capacity of a plant.

When the parametric equation is done, it is quick and easy to be used. Still, the equation elaboration demands time, historical data, and experts to define factors and execute the regression analysis. The technique is valuable and has the best accuracy range of all quick methods.

Figure 3.7 shows the steps to create the parametric equation.

Data acquisition fills a database with historical information listing features such as time, location, cost, power, capacity, and diameter. The features should be defined by an expert who knows the cost drivers for each type of project or equipment.

TABLE 3.6
Data – Exercise 4

Country	Factor	Cost Element Description	Cost (US$)
South Moon	4.10	Total equipment cost	41,500,000
North Sun	3.80		45,000,000

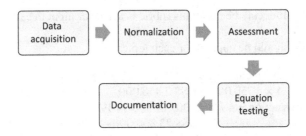

FIGURE 3.7   The process sequence of parametric equation creation.

The second step is data normalization, which consists of selecting an economic index and updating the cost reference for a specific date, as with Equation 3.5. Also, the data should be analyzed, and outliers should be deleted.

$$\text{Cost Updated} = \text{Original Cost} \times \frac{\text{Economic Index of the reference date}}{\text{Economic Index of the original date}}$$

(Equation 3.5)

The assessment step is to perform a regression analysis. In Excel, the data should be plotted through a scatter plot, clicking on the chart data with the right button, and the trendline should be selected, as shown in Figure 3.8. The choice is made by the higher R-squared (R2), which means a measure that represents how close the data is fitted in the regression model.

The equation test should verify the coherence and define the equation's application range. Figure 3.9 shows a group of cost references and the curve from the parametric equation (solid black curve). The hypothetical parametric equation based on power has a range of applications from 10 to 100 W. It occurs because it is unknown what will be the curve shape above 100 W. For example, it could be like the dashed line or double line curve in the Figure 3.9.

The equation should be reviewed periodically to check for market adherence, because inflation can generate an unrealistic result. Finally, the parametric equation should be registered, and all assumptions and data used for their creation.

3.5.1   Pump Cost Estimate – Example 9

Define the parametric equation and estimate the price for a 600 HP pump. Data from price pumps is listed in Table 3.7. The assumption is that all price references are hypothetical and have the same acquisition date, and pumps have the same material and requirements, except the power that varies like listed in the Table. Also, it is recommended that a minimum of 50 price references should be used, but for visualization purposes, only ten are used in the example.

Data acquisition is the first step. Then, the data normalization can be done through the scatter graph, like in Figure 3.10. The assumption is all price references are from equal dates. Then, the circle shows an outlier (pump 6 – 400 HP – $ 4,000.00). It

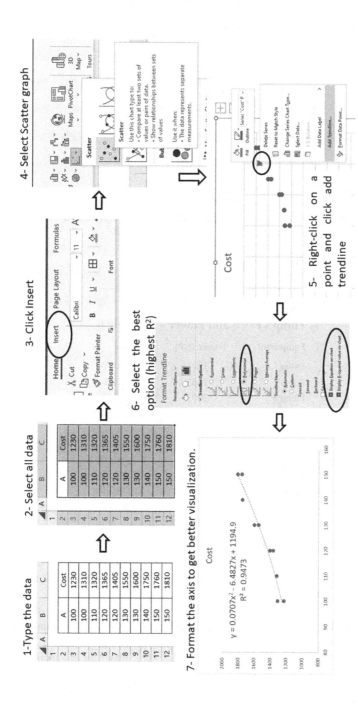

FIGURE 3.8   Step-by-step to add a trendline

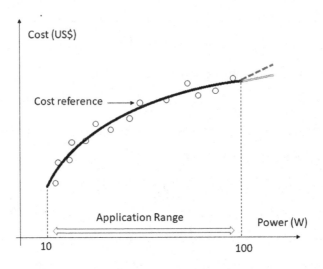

FIGURE 3.9 Application range of a parametric equation

TABLE 3.7
Pump's Price and Power – Example 9

Equipment	Power (HP)	Price
Pump 1	100	$ 1,000.00
Pump 2	150	$ 990.00
Pump 3	200	$ 1,200.00
Pump 4	220	$ 1,210.00
Pump 5	320	$ 1,450.00
Pump 6	400	$ 4,000.00
Pump 7	450	$ 2,400.00
Pump 8	480	$ 2,450.00
Pump 9	520	$ 3,000.00
Pump 10	650	$ 4,500.00

could be excluded because it can be an emergency acquisition or an exceptional condition that is not the reality in most cases. An expert should analyze the data to check for outliers in real life.

Next, performing a regression analysis is necessary, which is the assessment phase. The trend line could be defined in Excel by clicking on the chart data with the right button. Then, check the option to display the R2 and the equation. The equation choice should make the selection with greater R2, which in this example is the polynomial equation, like Figure 3.11.

The parametric equation $y = 0.0127\ x^2 - 3.2767x + 1247.7$ could be used to estimate the 600 HP pump cost.

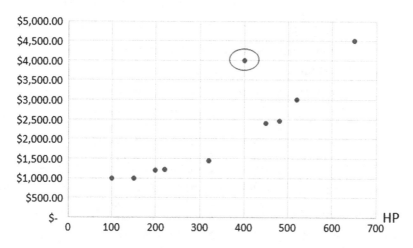

FIGURE 3.10   Scatter graph – Example 9

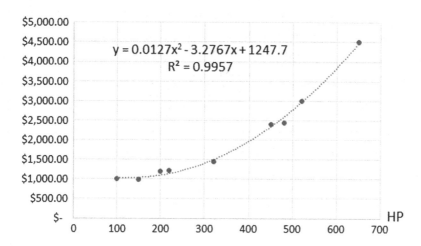

FIGURE 3.11   Scatter graph, trendline, and parametric equation – Example 9

Y = Cost estimate

Y = 0.0127 × (600)² − 3.2767 × (600) + 1247.7 = US$ 3,853.68. Using three significant figures, US$ 3.85 thousand.

3.5.2   Data Normalization – Example 10

Example 10 is similar to the last, but the pump acquisition date is different, like Table 3.8.

Hence, this example should update the prices to the same date, January 2022, during the normalization step. The hypothetical economic index used is listed in

TABLE 3.8

Pump's Price, Acquisition Date, and Power – Example 10

Equipment	Power (HP)	Price	Acquisition Date
Pump 1	100	$ 1,000.00	01 January 2021
Pump 2	150	$ 990.00	01 February 2021
Pump 3	200	$ 1,200.00	01 January 2021
Pump 4	220	$ 1,210.00	01 May 2021
Pump 5	320	$ 1,450.00	01 June 2021
Pump 6	400	$ 4,000.00	01 October 2021
Pump 7	450	$ 2,400.00	01 July 2021
Pump 8	480	$ 2,450.00	01 February 2021
Pump 9	520	$ 3,000.00	01 November 2021
Pump 10	650	$ 4,500.00	01 December 2021

TABLE 3.9

Economic Index – Example 10

Economic Index	
Date	Index
01-Jan-21	100.0
01-Feb-21	101.0
01-Mar-21	101.5
01-Apr-21	101.8
01-May-21	102.0
01-Jun-21	102.1
01-Jul-21	102.5
01-Aug-21	102.8
01-Sep-21	103.0
01-Oct-21	103.2
01-Nov-21	103.4
01-Dec-21	103.6
01-Jan-22	103.7

Table 3.9. The economic index should be coherent with country, industry, and material used in the real case.

Through Equation 3.5, the price is updated, e.g., pump 1:

$$\text{Price updated} = 1,000 \times \frac{100.0}{103.7} = \text{US\$ } 1,037$$

TABLE 3.10

Normalization – Example 10

DATA NORMALIZATION

Equipment	Power (HP)	Original Price	Updated Price
Pump 1	100	$ 1,000.00	$ 1,037.00
Pump 2	150	$ 990.00	$ 1,016.47
Pump 3	200	$ 1,200.00	$ 1,244.40
Pump 4	220	$ 1,210.00	$ 1,230.17
Pump 5	320	$ 1,450.00	$ 1,472.72
~~Pump 6~~			
Pump 7	450	$ 2,400.00	$ 2,428.10
Pump 8	480	$ 2,450.00	$ 2,515.50
Pump 9	520	$ 3,000.00	$ 3,008.70
Pump 10	650	$ 4,500.00	$ 4,504.34

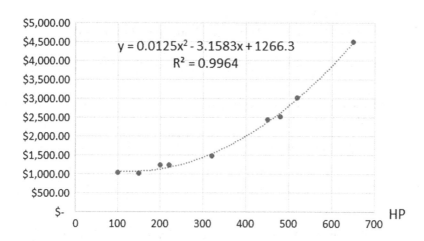

FIGURE 3.12 Scatter graph, trendline, and parametric equation – Example 10

The result after applying Equation 3.5 for all pumps is listed in Table 3.10. Also, pump 6 is excluded because it is an outlier, as discussed in the previous example.

Proceeding with the regression analysis by Excel, the polynomial equation is chosen because it has a higher R2, 0.9964. And the parametric equation is y = 0.0125x2 − 3.1583x + 1266.3, as shown in Figure 3.12.

Also, the 600 HP pump costs US$ 3871.32 or a difference of US$ 17.64 from the last example. If the significant figure method is observed, the cost is US$ 3.87 thousand. Despite the slight difference, if the normalization is not done, it can create considerable distortions and the result could become unrealistic.

3.5.3   OPEX Cost Estimate – Example 11

OPEX detailed in Section 2.10 can be estimated by a parametric equation. Equation 3.6 shows a hypothetical example of a parametric equation or OPEX estimation.

$$\text{OPEX cost} = \text{Cap} \times \text{RM} \times \text{OD} + \text{cap} \times \text{constant 1} + \text{MDay} \times \text{constant 2}$$

$$\text{(Equation 3.6)}$$

Where:

Cap = Capacity of the plant

RM = Raw material cost

OD = Operation days. It means how many days the plant is expected to operate per year.

MDay = Maintenance day. It means how many days the plant shut down for maintenance per year.

Table 3.11 provides data to exemplify Equation 3.6 for a process plant.

TABLE 3.11
Inputs for OPEX's Parametric Equation – Example 11

Data	Value	Unit
Raw material	1,000	US$/ton
Capacity	20	Ton/day
Operation days	330	Days
Maintenance days	30	Days
Constant 1	50,000	-
Constant 2	80,000	-

TABLE 3.12
Valve's Price, Acquisition Date, and Diameter – Exercise 5

Equipment	Diameter (in)	Price	Date
Valve 1	1.0	$ 10,000.00	01 January 2021
Valve 2	4.0	$ 20,500.00	01 February 2021
Valve 3	5.0	$ 25,550.00	01 March 2021
Valve 4	8.0	$ 30,000.00	01 May 2021
Valve 5	10.0	$ 35,000.00	01 June 2021
Valve 6	10.0	$ 70,000.00	01 February 2021
Valve 7	12.0	$ 41,000.00	01 July 2021
Valve 8	16.0	$ 48,000.00	01 February 2021
Valve 9	20.0	$ 55,000.00	01 November 2021
Valve 10	24.0	$ 80,000.00	01 October 2021

TABLE 3.13
Economic Index – Exercise 5

Economic Index

Date	Index
01-Jan-21	100.0
01-Feb-21	101.0
01-Mar-21	102.0
01-Apr-21	102.8
01-May-21	102.5
01-Jun-21	103.0
01-Jul-21	103.5
01-Aug-21	104.2
01-Sep-21	104.5
01-Oct-21	105.0
01-Nov-21	105.2
01-Dec-21	105.5
01-Jan-22	106.0

TABLE 3.14
Projects of Oil Ten Programme – Exercise 6

Project	Methodology	Data
FPSO A	Parametric equation	• Capacity new plant = 100,000 BOE per day • Cost = $500 \times$ Capacity $\times \log_2$(capacity) + 10^8
FPSO B	Parametric equation	• Capacity new plant = 150,000 BOE per day • Cost = $500 \times$ Capacity $\times \log_2$(capacity) + 10^8
Subsea pipeline	Physical dimensions	• Scope = 450 km • Cost per km = US$ 3.5 million
Oil terminal	Capacity factor	• Capacity new plant = 250,000 BOE per day • Cost similar plant (3 years ago) = US$ 200 million • Capacity similar plant = 300,000 BOE per day • Time factor (escalation) = 10% • Location factor = 5% • e = 0.65
Pipeline	Physical dimensions	• Scope = 50 km • Cost per km = US$ 3 million
Refinery	Capacity factor	• Capacity new plant = 250,000 BOE per day • Cost similar plant (3 years ago) = US$ 750 million • Capacity similar plant = 150,000 BOE per day • Time factor (inflation and escalation) = 14% • Location factor = –5% • E = 0.60

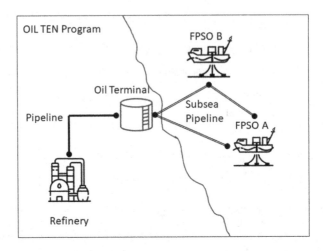

FIGURE 3.13 Oil ten programme scheme – exercise 6 and icons made by Freepick from www.flaticon.com

$$\text{Cost estimate} = 20 \times 1{,}000 \times 300 + 20 \times 50{,}000 + 30 \times 80{,}000$$
$$= \text{US\$ } 10{,}000{,}000$$

The example shows that the parametric equation has three variables, increasing the complexity considerably.

3.5.4 Parametric Equation – Exercise 5

Define the parametric equation for a valve whose cost is variable by diameter. Data is listed in Table 3.12 – the price reference should be updated to January 2022 by the economic index provided in Table 3.13.

What is the cost estimate for a 6" valve? Can the equation be used to estimate the cost of a 42" valve?

3.5.5 Program Cost Estimate – Exercise 6

The project manager asks you to develop a conceptual cost estimate for the Oil Ten Company Program, as in Figure 3.13, consisting of six projects listed in Table 3.14.

FPSO means floating production storage and offloading. It is a vessel used in the oil and gas industry. BOE means barrel of oil equivalent, a method to standardize a barrel of oil's energy.

Calculate each project cost and the total program cost. Each project should follow the method proposed in Table 3.14.

Disclaimer: The exercise does not consider all wells and subsea facilities. Also, the values and constants are hypothetical.

BIBLIOGRAPHY

ACostE, 2019. *Estimating Guide*. APM.

Fernando, J., 2021. *R-Squared Formula, Regression, and Interpretations*. Available at: https://www.investopedia.com/terms/r/r-squared.asp. [Accessed: 27 December 2022].

GAO, 2020. *Cost Estimating and Assessment Guide*. Available at: https://www.gao.gov/products/gao-20-195g. [Accessed: 15 December 2022].

Hastak, M., 2015. *Skills & Knowledge of Cost Engineering*, 6th Edition. AACE International.

NASA, 2015. *Cost Estimating Handbook*. Available at: https://www.nasa.gov/pdf/263676main_2008-NASA-Cost-Handbook-FINAL_v6.pdf. [Accessed: 21 January 2023].

EXERCISE ANSWERS

1- Process plant A – Green city – $ 2.0 billion. Process plan B – Big Sea – $ 1.7 billion, it has the lowest cost estimate.

2- Average – $ 8.9 million. Cheapest city – New Delhi – $ 2.2 million. Most expensive – Paris – $ 16.3 million.

3- Cost new plant – $ 8.4 million.

4- South Moon has the lowest cost, $ 170 million. North Sun – $ 171 million.

5- Parametric equation $y = 24.728x2 + 2053.7x + 12196$. And valve cost 42" is not recommended because there are no data with similar diameters; it extrapolates the application range.

6- FPSO A – $ 930 million. FPSO B – $ 1,390 million. Subsea pipeline – $ 1,575 million. Pipeline $ 150 million. Oil terminal – $ 205 million. Refinery – $ 1,104 million. Total program – $ 5,354 million.

4 Cost Estimate – Detailed Method

The detailed or bottom-up method quantifies and estimates each scope element. The model can use a deterministic or probabilistic approach.

The chapter focuses on the deterministic approach, and Chapter 5 discusses the probabilistic vision. The cost estimate is not an exact science because many uncertainties, such as quantity, planning, and cost reference variations, impact the result. Hence, the range of assumptions could promote different outcomes for the same scope.

Basically, the method consists of adding each cost component of labor, equipment, material, and other costs such as tax to achieve the project cost estimate, as shown in Figure 4.1.

Applications for the method are:

- Budget authorization
- Tenders
- Cost control
- Change orders

Although the method provides a better result when compared with the quick methods, it requires considerable time and resources to be implemented. It is directly related to the project complexity, as high complexity tends to increase the time and resources, as shown in Figure 4.2.

4.1 BASIS OF ESTIMATE (BOE)

BOE, according to AACE, is a document that defines the project's scope and ultimately becomes the basis for change management. When prepared correctly, any person with capital project experience can use the BOE to understand and assess the estimate, independent of any other supporting documentation.

The BOE is an essential component of every estimate, providing a straightforward narrative and supporting each decision point and the approach utilized. The benefits are listed below:

- Document the overall project scope
- Improve the review and validation of the cost estimate
- Provide a record of all documents used to prepare the estimate
- Alert potential cost risks and opportunities
- Record pertinent communications
- Source of information (e.g., for dispute resolution)

DOI: 10.1201/9781003402725-4

FIGURE 4.1 Bottom-up method and icons made by Freepick from www.flaticon.com

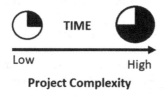

FIGURE 4.2 Execution time versus project complexity

FIGURE 4.3 BOE content and icons made by Freepick from www.flaticon.com

The BOE provides an approach for recording in a reliable and simple audit manner that supports an estimate's many types and applications.

BOE could be divided into four sections: design, planning, cost, and risk, as shown in Figure 4.3. Organizations may wish to vary this division to ensure the BOE supports each project delivery phase.

BOE should be organized and concise but include sufficient information to underpin the cost estimate. Also, it describes the estimating team, tools, methodology, and data used to develop the cost estimate. Furthermore, BOE must be transparent and aligned with the project scope.

Design basis includes information about the drawings, equipment list, and PID. Overall, it should detail the scope project from the client's requirement to the solution proposed. Estimation is a cyclic process which means, for example, after a design problem is detected, the design basis should be reviewed. Any discovery should be documented and captured in an updated estimate. This cyclic approach is applicable across all aspects of the estimate.

The design team is responsible for providing all relevant information to allow an estimate to be generated and for that estimate to be logically underpinned by the supporting documentation. The team responsible for the BOE must be updated if any document revision is undertaken. A different revision in the equipment list, for example, could affect the cost estimate if a pump or a heat exchanger is included or not.

Planning basis should inform the estimate about any productivity norms and assumptions included (e.g., overtime, location factors, etc.). They list the resources needed to complete each activity (e.g., how many trucks are necessary and how much time). The duration and sequence of the activities should be provided (e.g., schedule).

Cost basis provides all cost references, such as labor, subcontract, and equipment to be used in the estimate. In addition, it will inform the cost estimators, methodologies and tools used, as well as the communication done during the process. The date basis means the time reference for the cost estimate should be informed and allowances adopted.

Risk basis informs how contingency is defined and the main risks and opportunities that affect the cost estimate.

4.2   DETAILED METHOD – PROCESS

The process is summarized in Figure 4.4. It starts with the scope definition and take-off, and then the planning is elaborated. From this information, the cost estimator estimates the direct and indirect costs. Also, the cost estimate team should define the contingency considering the company's strategies and procedures. Then, the accuracy range should be estimated.

The asset or product price could be done by adding the overhead, interest rate, profit, and taxes to the cost estimate.

Along the process, the BOE is elaborated and reviewed, if necessary. After the cost estimate is concluded, an independent team should start the validation process and review the cost estimate.

FIGURE 4.4   Cost estimate process

4.3   SCOPE DEFINITION AND TAKE-OFF

The scope should be clearly defined because the result will not be a realistic value if it fails. It must be documented and assumptions registered. Also, any change along the cost estimate process should be recorded.

Techniques such as surveys, checklists, stakeholders meetings, and flowcharts, as in the example below, can help to verify if all scope is considered. Figure 4.5 shows a sequence to check all facilities for a process plant, from the plant itself through utilities, interconnections, safety installations, offices, and environmental requirements.

Then the take-off, which quantifies the material quantities associated with the project, is executed. Generally, it is supported by software that summarizes the amounts from the design, and each material should have a single identification or part number.

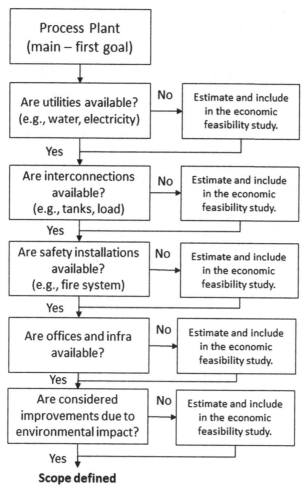

FIGURE 4.5   Example of scope definition sequence

4.4 PLANNING

The planning is based on the quantities from the take-off, assumptions, historical information, and company strategies. From the cost estimate perspective, it aims to estimate the necessary resources to implement the defined scope. It means determining who, when, and where each scope element will be executed. Hence, planning provides a detailed list of labor, material, and equipment used in the project. Also, the plan should result in a schedule, a master plan or a construction plan, and a planning plan reflecting the strategy, resources, and tools used during the scope execution.

Figure 4.6 shows a possible sequence to execute the planning for one material from the take-off, but it can differ according to each company's procedure. Consequently, the arrangement is applied for all materials listed on the take-off, resulting in a database with resources necessary to the proposed scope.

The process starts with the bill of materials and assumptions and constraints that should be analyzed and documented. Then, an assembly productivity index is defined for each material, and the crew responsible for the execution and the task duration are estimated. Finally, all the tasks are scheduled to create a master schedule, and a histogram with all resources is generated to support the cost estimation.

4.4.1 Assumptions and Constraints

The assumptions and constraints should be registered and analyzed. For example, they should discuss the weather and work conditions, work hours per day, overtime, and the calendar. To exemplify, the exact scope could have a different duration if it is done on the ground or at high height or if it is done in extreme weather (> 40°C or 0°C <).

4.4.2 Productivity Index

The productivity index (PI) is used to plan an activity, build a schedule, or evaluate the performance of a service. The index consists of the number of man-hours used to perform a given task.

As mentioned below, several factors must be assessed when defining or using a productivity index.

- Qualification of the workforce
- Team size
- Weather (e.g., rain, temperature, snow)
- Access
- Work conditions (e.g., work at height, confined space, etc.)

FIGURE 4.6   Example of planning sequence

PI for the same activity could be distinct if the task is done in a different environment. For example, welding a tube of the same material could have different PIs if the service is done in an open or confined space. Also, if the material size is different, the PI does not tend to be equal.

Equation 4.1 shows the productivity index formula.

$$PI = \frac{Mh}{AS}$$ (Equation 4.1)

Where:

PI = Productivity index

Mh = Man-hour

AS = Amount of service (e.g., ton, m, kg)

4.4.3 TYPICAL CREW

The typical crew should be defined according to historical information, software, or expert opinion. The team will be different per activity type and work conditions. For example, the crew responsible for building a wall varies by the material and equipment used. The team should be different if it is done in a modularized mode or a traditional one. Also, factors such as expertise and weather can affect the crew size.

Table 4.1 shows two possible crews for the pipe shop and assembly activity. However, the example is hypothetical because team composition and quantity could differ according to the technology and construction strategy used.

4.4.4 PRODUCTIVITY INDEX AND CREW – EXAMPLE 1

Company ABC needs to assemble a pipeline, and the client requests a deadline of 10 weeks. Considering the data and assumptions below, analyze if the time is feasible.

TABLE 4.1

Example of the Typical Crew for Pipe Shop and Assembly

Pipe shop crew	Quantity
Foreman	1
Welder	5
General helper	3
Pipe assembly crew	**Quantity**
Foreman	1
Welder	2
General helper	2
Plumber	2

Assumptions:

- Deadline = 10 weeks
- Quantity = 25 tons of 8" carbon steel
- Work time = 8 hours per day and 5 days per week
- No overtime
- PI for Pipe Assembly 8" – carbon steel, PI = 250 Mh/ton
- Crew available according to Table 4.2

The resolution is shown below. It is necessary to define how many man-hours (Mh) are required and compare them with Mh available, as in steps 1 and 2. Unfortunately, the man-hours available are not enough (step 3). Then, a new analysis is done considering overtime (+ 2 hours per day), as in steps 4 and 5. In this case, the result is satisfactory.

Step 1 – Total man-hours required through Equation 4.1 ➜ ***Mh*** = PI × AS

250 Mh/Ton × 25 Ton = 6,250 Mh

Step 2 – Total man-hours available

13 people × 8 hours × 5 days × 10 weeks = 5,200 Mh

Step 3 – Feasibility analysis ➜ Required × Available

6,250 > 5,200

Step 4 – New Mh available (with overtime)

13 people × 10 hours × 5 days × 10 weeks = 6,500 Mh

Step 5 – New feasibility analysis 1 ➜ Required × Available

6,250 < 6,500

TABLE 4.2
Typical Crew – Example 1

Category	Quantity
Welder	3
Plumber	4
General helper	5
Foreman	1
Total	13

4.4.5 ACTIVITIES DURATION

The activity duration is *"the length of time from start to finish of an activity, estimated or actual, in working or calendar time units"* (AACE International, 2022).[1] There are several methods to calculate it:

- Expert opinion
- Metrics
- Calculated. It is exemplified by the example below
- Project Evaluation and Review Technique (PERT)

Equation 4.2 shows the duration formula for the calculated method.

$$\textbf{Duration} = \frac{WQ}{AR}$$ (Equation 4.2)

Where:
WQ = Work quantity
AR = Applied resources

For example, the take-off lists one ton of eight inches of carbon steel to assemble a pipe. The company's historical information informs that the productivity index for this task is 200 man-hours per ton. Hence, 200 man-hours are necessary and considering that the crew has five professionals that work 8 hours per day, the duration is 5 days, as detailed below.

Duration = 1 ton × 200 man-hours per ton/(5 mans × 8 hours per day) =
Duration = 200 man-hours/40 man-hours per day = 5 days

4.4.6 TASK SCHEDULING

The planning should generate a schedule where all activities, durations, resources, and relationships are loaded. It informs the total time of the scope execution, critical path, and floats. Also, the resources from the schedule are one input for the cost estimate.

4.5 DIRECT AND INDIRECT COST ESTIMATES

The labor, material, and equipment resources should be cost-estimated through the data provided from the planning phase. Hence, the input is a resource list which describes each element, quantity, unit, duration, and additional information. It is called the Bill Of Materials (BOM) or Bill Of Quantities (BOQ).

Also, the cost database is essential information to ensure that each item has a cost reference used in the costing step. This database includes historical information, quotations, specialized sites or companies, and software. In addition, it must be updated periodically and the assumptions registered. The update means the economic indexes should normalize a past quotation to the current period.

One practice is that the company defines each element as a direct or indirect cost by the procedure. As mentioned in Chapter 2, the direct cost is related to the scope execution, which means these resources are clearly and adequately connected with the work. The indirect cost is not related to the activity, but it is necessary to ensure safety, quality, and management aspects, such as insurance.

The labor cost estimate is calculated considering the cost reference (e.g., salary), the category, quantity, fringe benefits, overtime, and work regime. The work regime is the hours each professional works per day or week. Also, additional costs, if necessary, should be analyzed and added, like admissions, training, personal safety equipment, and food costs.

The material and equipment cost estimate is defined from the technical features, description, quantity, unit, cost reference (e.g., quotation), delivery conditions, and any special requirements. Particular attention should be to ensure that all materials use the same currency, and the conversion rate applied should be documented. Equipment construction, such as trucks and cranes, should be verified if the operator/driver, maintenance, and fuel costs are included in the cost reference. Also, these machines should be determined if they are rented or a contractor's property. The depreciation should be calculated if the last option is adopted, as discussed in Chapter 9. Finally, the recommendation is that each material and equipment have a unique code that allows traceability, transparency, and organization.

Allowance is a percentage defined according to the discipline. It is defined *"as resources included in estimates to cover the cost of known but undefined requirements for an individual activity, work item, account or sub-account"* (AACE International, 2022).[2] This percentage refers to a part of the scope that is known but was not measured because there weren't sufficient details, or its quantification may not be economically efficient, such as screws and nuts. After the BOQ is estimated, the allowance could be added, as shown in Example 3.

After that, subcontracts can be included. Overall, they have the expertise and better cost than the hirer for specific services or activities. Quotations or historical information could be used to define the cost reference.

Insurance covers legal requirements, like life and contract insurance. Overall, the finance team provides the cost reference.

The warranty should be estimated based on the time and scope covered, and the historical information can help assess the cost or a percentage to be applied. It aims to cover repair and maintenance provided by the contractor during the warranty period.

In addition, owner costs could be added related to licenses, preliminary studies, the planning phase, project management, and contract management costs. Of course, these costs only are included when the owner does the estimation.

Figure 4.7 exemplifies a sequence to estimate the direct and indirect costs.

All assumptions, references, and historical information used should be documented. Also, the direct and indirect costs must be updated if the scope is modified along the cost estimate elaboration process.

4.5.1 ASSEMBLY UNIT

The assembly unit is a powerful method to simplify the costing process. It could be done only using labor, material, or a mix of labor and material inputs. The method saves time, promotes standardization, and traceability. Figure 4.8 exemplifies the method. It creates an assembly unit of 1 m² of the wall where labor and material are necessary. The unit cost is:

Labor cost = $ 8 + $ 4 = $ 12 and material cost = $ 7 + $ 2 + $ 4 = $ 13
Hence, assembly unit cost = $ 12 + $ 13 = $ 25

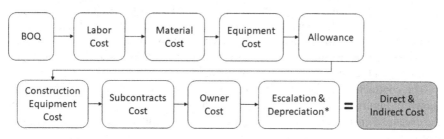

FIGURE 4.7 Direct and indirect cost estimate sequence

* **Factors such as depreciation and escalation should be included, but they are explained in Chapter 9.**

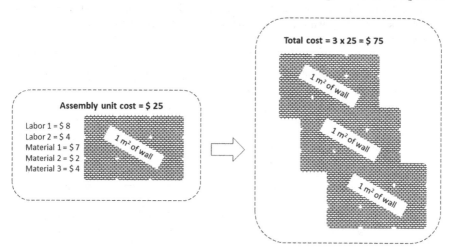

FIGURE 4.8 Assembly unit example

Then the assembly unit could be used for different areas during the cost estimate execution for each work package. Suppose the three square meters are required; then the cost estimate through the assembly unit is 3 × $ 25 = $ 75.

4.5.2 CONTINGENCY

Contingency aims to cover risks and uncertainties, as detailed in Chapter 5. There are several methods to estimate it, a fixed percentage, expert opinion, parametric equation, and risk analysis. The example in Section 4.11 adopted the fixed percentage, which should be based on a company procedure and can vary according to the project complexity and type.

4.6 PRICING, OVERHEAD, PROFIT, AND TAX ESTIMATES

As discussed in Chapter 2, this section covers the components that should be added to the cost estimate to create the price: overhead, interest rate, profit, and tax.

Overall, overhead should be adopted as a percentage. Hence, the per cent is applied to the total direct and indirect costs previously calculated. As mentioned in Chapter 2, the overhead aims to cover the legal, human resources, finance and accounting, IT, estimating, and sales departments.

The contractor executes the service and payments are made after a period, as shown in Figure 4.9. It means the company must have financial resources to cover the execution costs with a loan, for example. Hence, an interest rate could be applied to cover these costs.

Profit should be applied as a percentage of the total cost estimate (e.g., total cost estimate = direct + indirect cost + overhead + contingency). It is defined through business strategy, industry type, market conditions, demand, and project complexity. Consequently, the definition becomes a market strategy, and different percentages could be adopted for similar scenarios. For example, company A could adopt a low per cent because they want to gain a new market than company B, which has the expertise for the same service.

The profit could be added as a margin or markup. Despite both concepts being related to the price and having the same inputs, there are differences. Markup is applied to the cost through Equation 4.3:

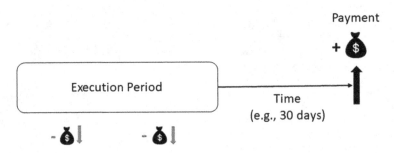

FIGURE 4.9 Payment example and icons made by Freepick from www.flaticon.com

$$\text{Markup} = C \times \text{Mk}\% \qquad \text{(Equation 4.3)}$$

Where:
C = Total cost
Mk% = Markup percentage

Also, the price could be estimated by Equation 4.4:

$$P = C \times (1 + \text{Mk}\%) \qquad \text{(Equation 4.4)}$$

Where:
C = Total cost
Mk% = Markup percentage
P = Price

Figure 4.10 left shows a markup example. 10% of the markup is applied in a total cost of $ 100. The result is a price of $ 110. Note that a 10% markup does not mean the price will have a 10% profit.

The margin is a percentage applied to revenue, and it is calculated through Equation 4.5. The price is estimated according to Equation 4.6.

$$\text{Margin} = P \times \text{Mg}\% \qquad \text{(Equation 4.5)}$$

Where:
P = Price
Mg% = Margin percentage

$$P = \frac{C}{\left(1 - \mathbf{Mg}\%\right)} \qquad \text{(Equation 4.6)}$$

Where:
C = Total cost
Mg% = Margin percentage
P = Price

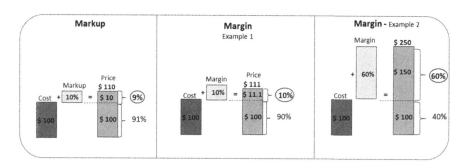

FIGURE 4.10   Markup (left) example and two margin (centre and right) examples

Using the same input of the markup example, Figure 4.10 (centre) shows that the price is different through the margin, like the calculation below:

$$P = \frac{100}{(1-10\%)} = \$111$$

And the margin is

$$\text{Margin} = 111 \times 10\% = \$\,11.1$$

The difference occurs because the margin is applied to the total (price), whereas the markup is applied to the cost. Also, Figure 4.10 (right) shows that if the margin is above 50%, the price becomes twice the cost or higher, like the calculation below:

$$P = \frac{100}{(1-60\%)} = \$250$$

$$\text{Margin} = 250 \times 60\% = \$\,150$$

Taxes should be analyzed according to current legislation, and each specific regulation should be known. One practice is including an accountant on the team to ensure that taxes are correctly considered and checking possible subsidies, for example.

4.7   BASELINE

The baseline is the

> budget that represents the approved scope of work and work plan. Identifiable plans, defined by databases approved by project management and client management, to achieve selected project objectives. It becomes the basis for measuring progress and performance and is the baseline for identifying cost deviations
>
> **(AACE International, 2022).**[3]

Hence, if the detailed (bottom-up) cost estimate is approved, it becomes the baseline. It is the sum of direct, indirect, contingency, and other costs (e.g., insurance), as shown in Figure 4.11. However, it can be different according to company procedures.

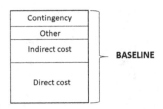

FIGURE 4.11   Baseline components

For example, the contingency could not be included in the project baseline but monitored separately.

The baseline should be documented, communicated, and controlled. It is expected not to be changed, but variations are common in real life. This discussion is done in Chapter 8.

4.8   ACCURACY RANGE

The deterministic cost estimate results in one number, but the accuracy range should accompany it, as shown in Figure 4.12. The accuracy range is *"an expression of an estimate's predicted closeness to final actual costs or time. Typically expressed as high/low percentages by which actual results will be over and under the estimate along with the confidence interval these percentages represent"* (AACE International, 2022).[4]

The accuracy is necessary because the cost estimate is not an exact science but a probabilistic one. Consequently, many factors, like risks and uncertainties, can affect the final project cost. Hence, the accuracy range provides a range within which the project is expected to end.

The range depends on the project's maturity. There are many uncertainties when the project is in the initial planning phase. Hence the accuracy range is higher than when the project has a detailed engineering design, as shown in Figure 4.13.

Each organization should have a procedure to define the correct range for each project's maturity. It should consider the cost estimate methodology, project complexity, and design maturity. Also, AACE International has several recommended practices that detail the accuracy range according to the industry type and project maturity, as listed in the bibliography section.

Figure 4.14 shows an example. The project has two planning phases. The initial planning phase has more uncertainties and risks than the second one. Hence, the accuracy range is higher (– 30%/+ 50%) than the second (– 15%/30%) when the project has a higher maturity. After the project ends is possible to detect that the final cost is under the accuracy range established along with the planning phase. Also, it is possible to see that contingency is added in the planning phases to cover these uncertainties and risks, and it assures that the project does not have a cost overrun.

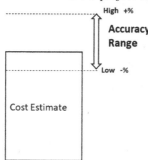

FIGURE 4.12   Accuracy range

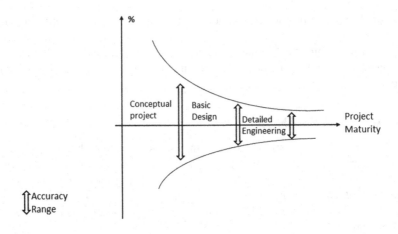

FIGURE 4.13   Accuracy range versus project maturity

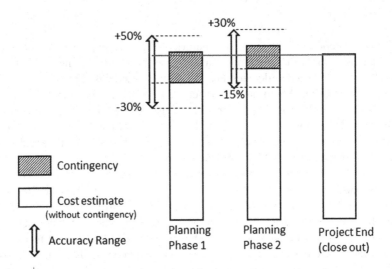

FIGURE 4.14   Accuracy range per phase.

4.9   COST ESTIMATE REVIEW

The cost estimate review aims to improve quality and reduce errors. Another objective is to evaluate whether the estimate is of sufficient quality to meet its intended purpose.

If possible, the revisor or the team responsible for the cost estimate revision should differ from the estimator team. In addition, the time required depends on the cost estimate and project maturity, which means that complex projects imply more time to revise. Also, senior professionals should be used because of their sensibility to find and check omissions and errors.

FIGURE 4.15 Pareto principle

It is highly recommended to use checklists to ensure all essential items are included and corrected. These could have different questions and topics according to project type. There is a list of possible topics below:

- Scope
- Basis of estimate (BOE)
- Assumptions
- Cost reference
- Direct/indirect cost
- Tax
- Profit
- Owners cost
- Engineering/design cost
- Escalation
- Contingency

When the time available is not enough, the Pareto Principle, as shown in Figure 4.15, could be a helpful tool because it focuses on essential items, for example, equipment and labor. The Pareto Principle states that for many outcomes, roughly 80% of consequences come from 20% of causes.

Table 4.3 shows the importance of cost estimate review to reduce error propagation. For example, the welder's wage was changed from $ 22 to $ 32, or a $ 10 variance because of a typing error. However, there are several multiplications to the labor cost estimate, as shown in Equation 2.1. Consequently, the final variance of this error is a difference of $ 3,600.

4.10 COST ESTIMATE VALIDATION

The difference between cost review and validation is that the first has a qualitative approach, and the second is quantitative. In addition, validation is crucial to ensure that strategic goals are achieved.

Also, the validation process occurs during or after the cost estimate review. According to the AACE International, 2022, validation is

a quality assurance process, typically quantitative in nature, to test or assure that an estimate of cost meets the project objectives and estimate cost strategy in regards to its appropriateness and purpose (which may include competitiveness or other organizational strategies identified for the estimate).

TABLE 4.3
Error Propagation Example

Category	Hourly wage ($)	Quantity	Hours per week	Fringe benefits (%)	Duration (weeks)	Total ($)
Welder	22 32	3	40	50	6	7,900
Welder		3	40	50	6	11,520

$\Delta = 10$ ———————————————————————————→ $\Delta = 3{,}600$

There are nine metric types to execute the cost validation:

- Cost/cost
- Hours/cost
- Cost/hours
- Cost/quantity
- Hours/hours
- Hours/quantity
- Quantity/quantity
- Quantity/cost
- Quantity/hours

As for the cost review, it is recommended that the cost validation team differs from the cost estimate team.

4.10.1   COST VALIDATION – EXAMPLE 2

Project XY has a cost estimate of $ 125 million and a 100 tons/day capacity. The cost breakdown is shown in Figure 4.16.

The cost validation is based on a successive detail method, which means that a group of key metrics are used initially. For the results that are not satisfactory, more metrics are added to understand inconsistencies.

Two indicators are used initially: total cost/capacity and direct cost (DC)/indirect cost (IC). The references are based on similar projects in the same industry.

Table 4.4 shows that the cost estimate is below one of the targets (total cost/capacity). However, the rate of DC/IC is not achieved, which could be an issue. Two additional metrics are analyzed to understand the cause: indirect cost/office hours and mechanical discipline cost/total ton of pipe.

The project should provide additional information to calculate these metrics:

Office hours estimated: 600,000 hours
Mechanical discipline cost: $ 15 million
Total ton of tubulation: 100 tons

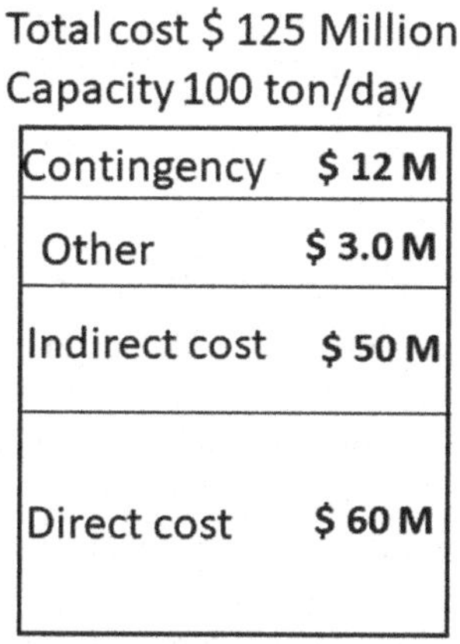

Total cost $ 125 Million
Capacity 100 ton/day

Contingency	$ 12 M
Other	$ 3.0 M
Indirect cost	$ 50 M
Direct cost	$ 60 M

FIGURE 4.16 Cost breakdown project XY

TABLE 4.4
Initials Metrics Project XY

Metric	Reference	Project XY result	Status
Total cost/capacity	< 1.6 $ million day/ ton	$ 125 million/100 ton/day = 1.3 $ million day/ton	 ✓
Direct cost/indirect cost	1.4 < Result < 1.6	$ 60 million/$ 50 million = 1.2	 😠

TABLE 4.5
Additional Metrics Project XY

Metric	Reference	Project XY result	Status
Indirect cost/office hours	< 60 $/man-hours	50 million/600,000 hours = 83 $/MH	 😠
Mechanical discipline: Cost/total ton of Pipe	0.2 < result < 0.3 $ million/ton	18 million/100 ton = 0.18 $ million/ton	 😠

The results of the additional metrics are shown in Table 4.5. The first metric shows $ 83 per man-hour above the 60 $/Mh target, indicating that indirect costs could be overestimated. The second indicator shows $ 0.18 million per ton of pipe below the expected range, highlighting a possible issue in the mechanical discipline. For example, the total quantity of pipes should be reviewed.

The indicators highlight that the cost estimate should be reviewed or justify this performance. Also, other metrics can be added to analyze the other disciplines, such as electrical, civil construction, etc.

4.11 COST ESTIMATE – DETAILED METHOD – EXAMPLE 3

The example aims to show an application of the detailed method, but there are simplifications to make it more didactic. For example, a bottom-up cost estimate could have thousands of items to be estimated. Still, it becomes unrealistic for this book's purpose, and a few items are listed in the example.

What is the cost estimate for the XYZ system described below? What is the price for the project?

Scope: Mechanical and electrical procurement, assembly, commissioning and start-up of the XYZ system. It consists of the equipment and instruments listed below and their interconnections, as shown in Figure 4.17.

- Vessel 01
- Pump 01
- Pump 02
- Tower 01
- Instruments P1 and L1

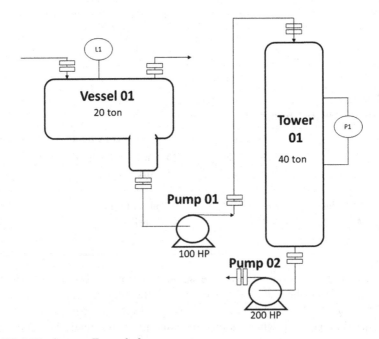

FIGURE 4.17   Scope – Example 3

Cost Estimate Inputs

Assumptions and Constraints

Planning
- Design and civil construction are executed before the work starts and are not included in the scope
- Sequence adopted according to the construction plan:
  - For equipment: Vessel 01 ➜ Tower 02 ➜ Pumps 01 and 02
- The pipe assembly and equipment assembly activities should start 2 weeks after the pipe shop activity starts
- Electrical and instrumentation start after equipment completion
- Electrical and instrumentation duration = 1 week
- Commissioning starts after equipment completion
- Commissioning duration = 3 weeks
- Start-up duration = 1 week and it starts after the commissioning,
- One typical lifting crew should be available for all service execution, except for the pipe shop activity. Assumption of four cranes and six trucks to support the activity
- Hours per week = 5 days × 8 hours per day = 40 hours
- Painting and insulation are subcontracted with 3 weeks of duration at the same time as commissioning

Costing and Pricing
- Escalation is zero (not considered)
- No overtime
- No owner cost because it is a cost estimate from the contractor's perspective
- Equipment and material quotations are provided as cost references
- Salary from a company database
- Subcontract – painting and insulation – $ 50 per m^2
- Miscellaneous = $ 5,000
- Allowances:
  - Tubulation allowance = 2%
  - Electrical and instrumentation allowance = 1.5%
  - Equipment allowance = 1%
- Office and facilities rate = $ 20,000 per month
- Warranty letter and construction insurances = $ 30,000
- Contingency = 20%. It is applied to direct and indirect costs
- Overhead = 2%. It is applied to direct and indirect costs
- Interest rate = 2%. It is applied to direct and indirect costs
- Profit (margin) = 15%
- Tax = 20% of price (percentage of total)

Quantities from Take-off
- Piping to be assembled = 20 ton
- Piping to be manufactured at pipe shop = 10 ton
- Vessel 01 weight = 20 ton

- Tower 01 weight = 40 ton
- Pump 1 = 100 HP
- Pump 2 = 200 HP
- Painting and insulation = 1,000 m^2

Productivity Index

- Pipe assembly = 275 Mh/ton (man-hour per ton). The assumption is that all material is the same type and diameter, and the assembly complexity is always equal.
- Pipe shop = 180 Mh/ton. The assumption is that all material is the same type and diameter, and the complexity is always equal.
- Vessel 01 assembly = 50 Mh/ton
- Tower 01 assembly = 40 Mh/ton
- Pump 01 and 02 = 3 Mh/HP

Crew – Direct Labor

- Fringe benefits = 58%. It is included food, commuting, health and life insurance, and other benefits
- The direct labor is listed in Table 4.6 per discipline.

TABLE 4.6
Crews of Direct Labor – Example 3 – The team is hypothetical and does not necessarily reflect a real team

Pipe shop	Quantity	Pipe assembly	Quantity
Foreman	2	Foreman	4
Welder	10	Welder	8
General helper	6	General helper	8
Total	18	Plumber	8
		Total	28
Equipment assembly	**Quantity**	**Eletrical and instrumentation**	**Quantity**
Foreman	3	Electrical technician	1
General helper	6	Instrumentation technician	1
Mechanical assembler	15	Total	2
Total	24		
Commissioning and start-up	**Quantity**	**Lifting**	**Quantity**
Commissioning specialist	2	Driver crane	1
Foreman	1	Lifting technician	1
Welder	1	General helper	1
General helper	1	Truck driver	1
Plumber	1	Total	4
Mechanical assembler	1		
Electrical technician	1		
Instrumentation technician	1		
Total	9		

Crew – Indirect Labor

- Fringe benefits = 48%. It is included food, commuting, health and life insurance, and other benefits
- The indirect labor is listed in Table 4.7.

Indirect labor should be constant for all contract duration. Also, the design team is for clarifications, change control, and as-built.

Equipment and Material Quotation
The cost reference, listed in table 4.8, has met all requirements and specifications, and the delivery is on-site. Also, there is an allowance detailed per material.

Construction Equipment and Additional Items
Table 4.9 lists the construction equipments and additional items.
    The resolution is divided into planning, costing, and pricing.

TABLE 4.7
Indirect Labor – Example 3 – The team is hypothetical and does not necessarily reflect a real team

Project Management Team	Quantity	Design Team	Quantity
Project manager	1	Engineer	2
Planner	1	Contract management team	
Quantity surveyor	1	Manager	1
Quality and safety team		Administrator	1
Engineer	1	Secretary	1
Technician	2	Warehouse team	
		Warehouse officer	1

TABLE 4.8
Equipment, Material Quotation, and Allowance

Item	Cost Reference	Unit	Allowance
Pump 01	40,000	$/unit	1.0%
Pump 02	50,000	$/unit	1.0%
Vessel 01	120,000	$/unit	1.0%
Tower 01	150,000	$/unit	1.0%
Carbon steel pipe 8" (20 ton – lump sum)	40,000	$	2.0%
Miscellaneous (lump sum)	5,000	$	
Electrical and instrumentation (lump sum)	10,000	$	1.5%

TABLE 4.9

Construction Equipment and Safety Equipments

Item	Cost Reference	Unit	Quantity
Crane	6,000	$/month	2
Truck	1,500	$/month	3
Tools	100	$/man	85
Safety equipments	100	$/man	100

4.11.1 PLANNING RESOLUTION – EXAMPLE 3

Following the Figure 4.6 sequence, the material provided by take-off is multiplied by the productivity index to obtain the total man-hours, like below:

- Piping to be manufactory at pipe shop = 10 ton × 180 Mh/ton = 1,800 Mh
- Tubulation to be assembly = 20 ton × 275 Mh/ton = 5,500 Mh
- Vessel 01 height = 20 ton × 50 Mh/ton = 1,000 Mh
- Tower 01 height = 40 ton × 40 Mh/ton = 1,600 Mh
- Pump 01 = 100 HP × 3 Mh/HP = 300 Mh
- Pump 02 = 200 HP × 3 Mh/HP = 600 Mh

After that, the typical crew listed in Table 4.6 could be used to estimate the duration of each activity. The total of each crew is multiplied by hours per week from the assumptions:

- Pipe shop duration = 1,800 Mh/(18 Mans × 40 hours per week) = 1,800/720 = 2.5 weeks
- Pipe assembly duration = 5,500 Mh/(28 Mans × 40 hours per week) = 5,500/1,120 = 4.9 weeks
- Vessel 01 assembly duration = 1,000 Mh/(24 Mans × 40 hours per week) = 1,000/960 = 1 week
- Tower 01 assembly duration = 1,600 Mh/(24 Mans × 40 hours per week) = 1,600/960 = 1.7 weeks
- Pump 01 assembly duration = 300 Mh/(24 Mans × 40 hours per week) = 300/960 = 0.3 week
- Pump 02 assembly duration = 600 Mh/(24 Mans × 40 hours per week) = 600/960 = 0.6 week

The assumptions and constraints section informs the construction plan's sequence and the additional activity's duration.

- Sequence adopted according to the construction plan:
  - For equipment: Vessel 01 ➔ Tower 02 ➔Pumps 01 and 02

- The pipe assembly and equipment assembly activities should start 2 weeks after the pipe shop activity starts
- Electrical and instrumentation start after equipment completion and the duration is 1 week
- Commissioning starts after equipment completion
- Commissioning duration = 3 weeks
- Start-up duration = 1 week

Finally, as a simplification, pumps 01 and 02 are assembled at the same week, and all activities' duration is rounded up. The task scheduling results in the schedule shown in Figure 4.18.

Hence, the scope deadline is 10 weeks or 2.5 months. In addition, the histogram is shown in Figure 4.19, and it is helpful to estimate the facilities, such as restaurants, offices, and transportation. Also, they are critical to the next step, the cost estimate, because they inform the total of direct labor resources. For example, the pipe show activity has 18 professionals per week on the crew. So the number 18 appears each week of the histogram in the first line.

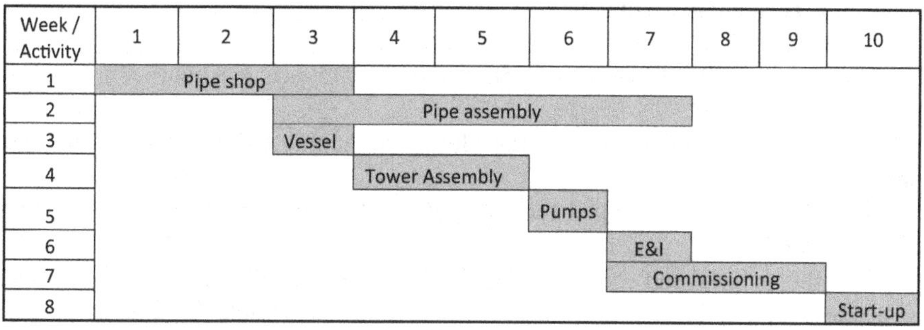

FIGURE 4.18 Schedule – Example 3

Week / Activity	1	2	3	4	5	6	7	8	9	10	
1	18	18	18								
2			28	28	28	28	28				
3			24								
4				24	24						
5						24					
6							2				
7							9	9	9		
8										9	
Sub total	18	18	70	52	52	52	39	9	9	9	1640
Lifting team			4	4	4	4	4	4	4	4	160
Total											**1800**

The total number of professionals per week and activity is shown here. It comes from table 4.6.

Subtotal is the weekly sum multiplied by 5 days.

FIGURE 4.19 Histogram – Example 3

4.11.2   Cost Estimate Resolution – Example 3

The sequence adopted is direct labor, material and equipment, construction equipment and additional items, subcontract cost, indirect labor, office and facilities, warranty and insurance.

The direct labor cost is estimated by Equation 2.1:

$$LC = Salary \times PQ \times (1 + FB) \times (1 + OT) \times Duration$$

Where:
LC = Labor cost
PN = Professional's quantity
FB = Fringe benefits
OT = Overtime

Planning provides that the crew, quantity, and duration. Fringe benefits are 58%. The assumptions show that there is no overtime and the salary is from the company's database, as shown in Table 4.10. For example, the foreman cost estimate of the pipe show crew is calculated as below.

Foreman cost = $ 24 × 2 × (1 + 0.58) × (1 + 0) × 15 days × 8 hours = $ 9,101

As the pipe shop activity has 3 weeks duration meaning 15 days × 8 hours or 120 hours.

The same formula is applied for each category, resulting in a total direct labor cost of $ 372,438.

Then, the material and equipment (ME) could be estimated. The example is based on a short list, and the cost reference is from a quote, as in Table 4.11 below.

An allowance should be added to cover the cost of known but undefined requirements. For example, the pump cost is:

Pump cost reference × allowance % = 40,000 × (1 + 0.01) = $ 40,400. This process is done for each item in Table 4.11.

The total cost for materials and equipment is $ 419,550.

After that, construction equipment (CE) and additional items are estimated, as listed in Table 4.12. The CE is calculated considering they are rented, including maintenance and operation costs. However, the drivers are not included in the cost per month because they were estimated in the lifting crew.

The number of necessary tools is 85, the same number of direct labor from Table 4.6. The safety equipment quantity is the total of direct and indirect labor, which is 100 from Tables 4.6 and 4.7.

The estimate is obtained by multiplying the cost by the quantity. Also, the duration, 2.5 months, should be included for crane and truck costs. For example, the crane cost is $ 6,000 × 2 cranes × 2.5 months = $ 30,000.

TABLE 4.10
Direct Labor Cost

Category	Hourly base wage ($)	Quantity	Duration (days)	Cost ($)
Pipe shop crew				61,051
Foreman	24	2	15	9,101
Welder	19	10	15	36,024
General helper	14	6	15	15,926
Pipe assembly crew				159,264
Foreman	24	4	25	30,336
Welder	19	8	25	48,032
General helper	14	8	25	35,392
Plumber	18	8	25	45,504
Equipment assembly crew				107,693
Foreman	24	3	20	18,202
General helper	14	6	20	21,235
Mechanical assembler	18	15	20	68,256
Electrical and instrumentation crew				2,970
Electrical technician	23	1	5	1,454
Instrumentation technician	24	1	5	1,517
Commissioning and start-up crew				53,341
Commissioning specialist	34	2	20	17,190
Foreman	24	1	20	6,067
Welder	19	1	20	4,803
General helper	14	1	20	3,539
Plumber	18	1	20	4,550
Mechanical assembler	18	1	20	4,550
Electrical technician	25	1	20	6,320
Instrumentation technician	25	1	20	6,320
Lifting crew				41,459
Driver crane	26	1	40	13,146
Lifting technician	24	1	40	12,134
General helper	14	1	40	7,078
Truck driver	18	1	40	9,101
Total labor direct cost				372,438

Then, the painting and insulationsubcontract is estimated as below.

$$\text{Subcontract cost} = 1{,}000 \text{ m}^2 \times 50 \text{ \$/m}^2 = \$\ 50{,}000$$

Next, the indirect cost is estimated. Formula 2.1 is used for the indirect labor cost, but there is a critical difference from the direct cost estimate. The salary is monthly,

TABLE 4.11
Material and Equipment

Material and Equipment (ME)				Cost ($)
Item	Cost Reference	Unit	Allowance (%)	
Pump 1	40,000	$/unit	1	40,400
Pump 2	50,000	$/unit	1	50,500
Vessel	120,000	$/unit	1	121,200
Tower	150,000	$/unit	1	151,500
Carbon steel pipe 8" (20 ton)	40,000	$	2	40,800
Miscellaneous	5,000	$	0	5,000
Electrical and instrumentation	10,000	$	1.5	10,150
Total ME cost				419,550

TABLE 4.12
Construction Equipment and Additional Items

Construction Equipment and Additional Items (CE)				CE cost ($)
Item	Cost	Unit	Quantity	
Crane	6,000	$/month	2	30,000
Truck	1,500	$/month	3	11,250
Tools	100	$/man	85	8,500
Safety equipments	100	$/man	100	10,000
Total CE cost				59,750

and the duration should be the same, a monthly period. Also, the fringe benefits are 48%. For example, the project manager's cost estimate is:

Project manager cost = 6,000 × 1 × (1 + 0.48) × (1 + 0) × 2.5 months = $ 22,200.

The same formula is applied for each category, resulting in a total of $ 247,900, as listed in Table 4.13.

The office and facilities, such as the warehouse, toilets, and restaurant, are estimated as rated per month:

Rate = $ 20,000 per month
Duration = 2.5 months
Office and facilities cost = $ 20,000 × 2.5 months = $ 50,000
The warranty letter and insurances have a cost of $ 30,000

TABLE 4.13
Indirect Labor Cost

Category	Monthly base wage ($)	Quantity	Duration (months)	Cost ($)
Project management team				61,050
Project manager	6,000	1	2.5	22,200
Planner	5,500	1	2.5	20,350
Quantity survey	5,000	1	2.5	18,500
Quality and safety team				56,240
Engineer	7,000	1	2.5	25,900
Technician	4,100	2	2.5	30,340
Design team				51,800
Engineer	7,000	2	2.5	51,800
Contract management team				63,270
Manager	9,000	1	2.5	33,300
Administrator	6,000	1	2.5	22,200
Secretary	2,100	1	2.5	7,770
Warehouse team				15,540
Warehouse office	2,100	2	2.5	15,540
Total labor indirect cost				247,900

Hence, the direct and indirect costs can be summarized below:

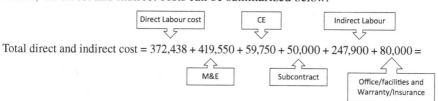

Total direct and indirect cost = 372,438 + 419,550 + 59,750 + 50,000 + 247,900 + 80,000 =

Total direct and indirect cost = $ 1,229,638

The next step is adding contingency through the percentage defined in the assumptions section.

Contingency: 20%
Contingency applied to DC + IC = 20% × $ 1,229,638 = $ 245,928
Hence the cost estimate baseline is $ 1,229,638 + $ 245,928 = $ 1,475,565.

Finally, the accuracy range should be estimated. There are many simplifications, and considering the project maturity is not high, the range adopted is − 10/+ 30% or a Class 3 according to the AACE RP 18R-97, like Figure 4.20. Hence, the final cost (or actuals) is expected to be between $ 1,106,674 and $ 1,598,529. Also, Figure 4.20 shows the project baseline compound by the direct, indirect costs and contingency.

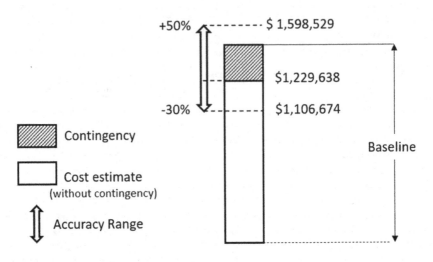

FIGURE 4.20 Accuracy range – Example 3

4.11.3 PRICING RESOLUTION – EXAMPLE 3

Price is obtained by adding overhead, interest rate, profit, and taxes. These percentages come from the assumptions section.

All these percentages are defined according to the company procedure, and the profit should consider the company's strategy for each project.

Overhead rate: 2%. It is applied to direct and indirect costs.

$$\text{Overhead cost applied to DC + IC} = 2\% \times \$\ 1{,}229{,}638 = \$\ 24{,}593$$

Interest rate: 2%. It aims to cover the spending money from loans. For simplification, it is adopted a fixed percentage is applied to the direct and indirect costs.

$$\text{Interest cost applied to DC + IC} = 2\% \times \$\ 1{,}229{,}638 = \$\ 24{,}593$$

Profit: 15%. It is calculated as a margin.

Tax: 20%. It is calculated as a percentage of the total, which means that the formula is similar to the margin concept.

$$\text{Profit} = \frac{(\text{Cost estimate baseline} + \text{overhead} + \text{interest rate cost})}{(1 - \text{margin}\% - \text{Tax}\%)} \times \text{margin}\%$$

$$\text{Profit} = \frac{(1{,}475{,}565 + 24{,}593 + 24{,}593)}{(1 - 15\% - 20\%)} \times 15\% = \$\ 351{,}866$$

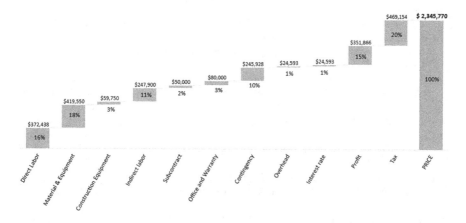

FIGURE 4.21 Price breakdown

$$\text{Tax} = \frac{(\text{Cost estimate baseline} + \text{overhead} + \text{interest rate cost})}{\left(1 - \text{margin}\% - \text{tax}\%\right)} \times \text{tax}\%$$

$$\text{tax} = \frac{\left(1,475,565 + 24,593 + 24,593\right)}{\left(1 - 15\% - 20\%\right)} \times 20\% = \$\,469,154$$

Hence, the price is:

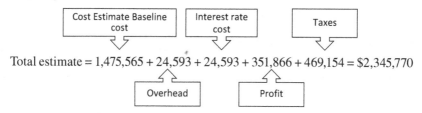

$$\text{Total estimate} = 1,475,565 + 24,593 + 24,593 + 351,866 + 469,154 = \$2,345,770$$

Figure 4.21 provides a waterfall graph, where each component and its respective percentage illustrate the price breakdown. Contingency, overheard, and interest rates have different percentages than the assumption section, 10%, 1%, and 1%, respectively, because the breakdown is related to the price, not direct and indirect costs.

NOTES

1-4. Reprinted with permission from AACE International. Check the website for the latest versions (https://web.aacei.org/resources/cost-engineering-terminology).

BIBLIOGRAPHY

AACE International, 2020a. *Recommended Practice 17R-97 Cost Estimate Classification System.* Available at: https://web.aacei.org/resources/recommended-practices. [Accessed: 11 February 2023].

AACE International, 2020b. *Recommended Practice 18R-97 Cost Estimate Classification System – As Applied in Engineering, Procurement, and Construction for the Process Industries.* Available at: https://web.aacei.org/resources/recommended-practices. [Accessed: 11 February 2023].

AACE International, 2020c. *Recommended Practice 100R-20 Cost Estimate Validation.* Available at: https://web.aacei.org/resources/recommended-practices. [Accessed: 2 February 2023].

AACE International, 2022. *Recommended Practice 10S-90 Cost Engineering Terminology.* Available at: https://web.aacei.org/resources/cost-engineering-terminology. [Accessed: 8 January 2023].

ACostE, 2019. *Estimating Guide.* APM.

GAO, 2020. Cost Estimating and Assessment Guide. Available at: https://www.gao.gov/products/gao-20-195g. [Accessed: 15 December 2022].

Hastak, M., 2015. *Skills & Knowledge of Cost Engineering*, 6th Edition. AACE International.

Investopedia Team, 2021. *Profit Margin vs. Markup: What's the Difference?.* Available at: https://www.investopedia.com/ask/answers/102714/whats-difference-between-profit-margin-and-markup.asp#:~:text=The%20profit%20margin%2C%20stated%20as,arrive%20at%20the%20selling%20price. [Accessed: 5 February 2023].

NASA, 2015. *Cost Estimating Handbook.* Available at: https://www.nasa.gov/pdf/263676main_2008-NASA-Cost-Handbook-FINAL_v6.pdf. [Accessed: 21 January 2023].

5 Probabilistic Cost Estimate

The previous chapter discussed the deterministic cost estimate, which results in a number followed by an accuracy range. It occurs because cost estimate is not an exact science. Still, it generates a range of values because uncertainties and risks are associated with each project's elaboration process.

Hence, the probabilistic approach is discussed in this chapter because it can create a strong result once risks and uncertainties are considered.

Uncertainties are related to planning, quantities, cost reference variations, and methods adopted. Risks, according to (PMI, 2021 2013),

> is an uncertain event or condition that, if it occurs, has a positive or negative effect on one or more project objectives such as scope, schedule, cost, and quality. A risk may have one or more causes and, if it occurs, it may have one or more impacts. A cause may be a given or potential requirement, assumption, constraint, or condition that creates the possibility of negative or positive outcomes.

The probabilistic cost estimate is created by understanding and measuring the uncertainties and risks, and it results in probabilistic scenarios, whereas the contingency is estimated for the project vision.

A different approach could be adopted if the portfolio is used instead of the project vision. In this case, the management reserve is estimated to protect the portfolio.

5.1 PROBABILISTIC COST ESTIMATE DUE TO UNCERTAINTIES

The deterministic cost estimate is based on a specific value or cost reference for each scope component, labor, material, and equipment. However, there are uncertainties related to quantities, planning, cost references, and cost estimation methodology. For this reason, the probabilistic cost estimate is elaborated to include these uncertainties defining a range of possible values for them.

Each method has uncertainties related to it, as illustrated in Figure 5.1. An expert's opinion has an uncertain high level because it is based exclusively on experience. The analogy technique discussed in Chapter 3 allows for analyzing only a few specific characteristics of each project because it seeks to be a fast method with low accuracy. The parametric method can consider several factors, reducing the level of uncertainties. Still, it does not overcome the detailed method where each scope element, labor, material, and equipment are considered, as seen in Chapter 4.

Other possible variations are exemplified in Figure 5.2. The left image shows that each labor cost could have uncertainties related to the salary, for example. Overall,

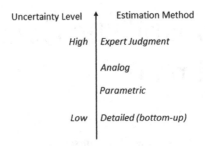

FIGURE 5.1 Uncertainty level related to the estimation method

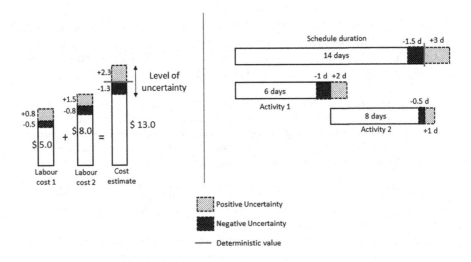

FIGURE 5.2 Examples of productivity and cost reference variations

the cost reference uncertainties are related with factors as inflation, escalation, and market conditions. Hence, the cost estimate has a level of uncertainty leading that the final cost could vary from $ 11.7 to $ 15.3, as shown below:

$$\text{Worst scenario} = 5 + 0.8 + 8 + 1.5 = \$ 15.3$$

$$\text{Best scenario} = 5 - 0.5 + 8 - 0.8 = \$ 11.7$$

The right image shows that the schedule could have uncertainties related to productivity. It could vary because of learning curves, training, weather conditions, etc. Both activity durations can change, resulting in a final time between 12.5 and 17 days.

$$\text{Worst scenario} = 6 + 2 + 8 + 1 = 17.0 \text{ days}$$

$$\text{Best scenario} = 6 - 1 + 8 - 0.5 = 12.5 \text{ days}$$

As a result, the probabilistic cost estimate uses multiple scenarios to embrace these possibilities, where each component has a different value per simulation. The distribution probability types define the range of possible values for scope elements, and the most common types are normal, triangular, uniform, lognormal, beta, poison, and Bernoulli.

The book focuses on triangular probability distribution to illustrate the probability vision of the cost estimate. The triangular distribution is based on three points: most likely, pessimistic, and optimistic values, like in Figure 5.3.

The proposed method for the probabilistic cost due to the uncertainties is shown in the sequence in Figure 5.4. The deterministic cost estimate discussed in the previous chapter is the starting point. Then, factors affected by variations should be analyzed, and the triangular distribution should define a range of possible values for each uncertainty. Then, the Monte Carlo simulation is run, and the probabilistic cost is estimated through the cumulative probability curve.

One feature of the uncertainties is that they increase or decrease according to the project's maturity. When the project starts and maturity is low, the level of uncertainties is high because many definitions should be done related to scope, technology, strategies, etc. Also, the opposite occurs when planning and design are executed, the maturity increases, and the uncertainties become low.

5.1.1 Probabilistic Cost Estimate Due to Uncertainties – Example 1

The example is based on Section 2.1, where Company ABC estimates the cost of assembling pipes. The deterministic result, $ 88,560, is shown in Table 5.1 and calculated through Equation 2.1.

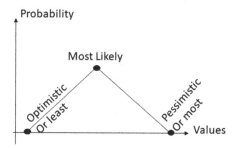

FIGURE 5.3   Triangular distribution

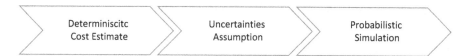

FIGURE 5.4   Contingency estimation sequence due to uncertainties.

However, Company ABC has uncertainties related to salary and duration. They defined a range of possible wages and time based on similar projects through a triangular distribution. Table 5.2 shows the inputs for each hourly salary and duration.

The next step is to generate the probabilistic cost estimate through the Monte Carlo simulation. It is defined that 5,000 simulations will be created, meaning that after running the simulation, 5,000 distinct cost estimates are calculated. Each interaction has a different salary for team members and duration to estimate the cost by Equation 2.1.

The statistic result is shown in Table 5.3. The range of possible values is from $ 62,317, minimum, to $ 134,055, maximum. It shows that many simulations have a cost above the deterministic cost estimate. It is corroborated by the average of $ 93,417, approximately five thousand over the deterministic cost. Also, the percentiles show that around 40% of interactions are under the initial cost estimate.

Figure 5.5 shows the histogram, and the solid curve called the cumulative probability curve. The columns show the frequency at which the result appears during the simulation. For example, $83,000 performs around 10% of the interactions or 500 times. The cumulative probability curve adds the outcome of each interaction.

The dashed line shows that the deterministic cost estimate, $ 88,560, covers 41% of the simulations. Hence, 59% of possible results promote a cost overrun when the execution cost exceeds the deterministic cost. Also, It is important to mention that no risk has been considered until now, just the uncertainties.

TABLE 5.1

Deterministic Labor Cost Estimate from Section 2.1

Category	Hourly salary ($)	Quantity	Hours per week	Fringe benefit (%)	Duration (weeks)	Total ($)
Welder	22	3	40	0.5	6	23,760
Plumber	20	4	40	0.5	6	28,800
General helper	15	5	40	0.5	6	27,000
Foreman	25	1	40	0.5	6	9,000
						88,560

TABLE 5.2

Inputs of Triangular Distribution for Probabilistic Cost Estimate – Example 1

Uncertainties inputs	Welder ($)	Plumber ($)	General helper ($)	Foreman ($)	Duration (weeks)
Lowest (worst)	20	17	12	20	5
Most likely	22	20	15	25	6
Highest (best)	24	23	19	28	8

TABLE 5.3
Statistics Result – Example 1

Deterministic cost estimate ($)	88,560
Maximum ($)	134,055
Minimum ($)	62,317
Average ($)	93,417
Median ($)	92,088
Percentiles	
10	74,113
25	82,046
40	88,500
50	92,088
75	104,170
90	114,588

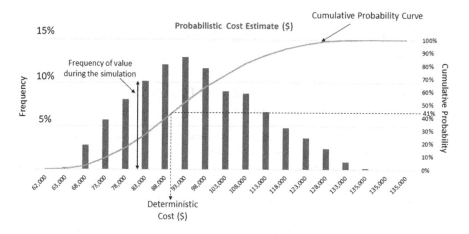

FIGURE 5.5   Probabilistic cost estimate due to uncertainties – histogram of frequencies, columns, cumulative probability curve, orange curve

5.2   RISK ASSESSMENT – PROBABILISTIC COST ESTIMATE DUE TO RISKS AND UNCERTAINTIES

Although uncertainties promote a realistic scenario for each cost estimate, risks must be included for a complete analysis. Hence, the probabilistic cost estimate should consider the risk assessment. The sequence in Figure 5.4 is revised to input the risks and uncertainties, as shown in Figure 5.6.

Risk management is not the book's focus, but some aspects should be discussed. Risks are analyzed in the risk identification and assessment shown in Figure 5.6.

FIGURE 5.6   Probabilistic cost estimation sequence due to uncertainties and risks

Risk identification aims to identify the risks that can affect the project, and they could be positive, as an opportunity, or negative, as a threat. After the identification stage, they should be assessed through the likelihood of occurring and impact.

The impact could be measured in several ways, such as triangular, normal, Bernoulli, uniform, and lognormal distributions.

After identifying and assessing the risks, the probabilistic simulation generates the probabilistic cost estimate. Furthermore, running and creating a second or more simulations is possible. It occurs because the risk needs to be treated with mitigation plans or even eliminated, for example, with a scope change. Hence, others scenarios could be created considering the mitigation actions, reducing the impact or probability of each risk.

The point choice in the cumulative probability curve to be the project cost estimate is based on market and business conditions. Ultimately, it is based on manager/board experience.

5.2.1   Risk Assessment – Probabilistic Cost Estimate Due to Uncertainties and Risks – Example 2

Example 1 only considers uncertainties, but real cases should include risks and uncertainties. Hence, Company ABC evaluated the pipe assembly project following the process proposed in Figure 5.7.

The two first steps were done in Example 1, and the risk identification identifies three risks, as listed in Table 5.4. The new technology is based on new assembly tools that could improve productivity, but because it will be the first time using it. Hence the likelihood is low, 30%. The streak is a threat with a medium probability of 50%. Finally, there is a threat of work permission delay and a high likelihood of 75%.

The triangular distribution measures the risk impact, where three points are estimated. The new technology is an opportunity but has a negative effect because it reduces the planned amount. Threats have a positive impact because they are cost increases. Table 5.5 summarizes the results per risk.

After that, the probabilistic cost estimate is elaborated through 5,000 interactions. An aleatoric value for uncertainties is used in Equation 2.1 in each scenario, as discussed in Example 1. In addition, a different value through triangular distribution is adopted when a risk occurs following the likelihood chosen. Figure 5.8 shows the process for each simulation — for example, in interaction one, the salary and duration decrease. Also, risks two and three occur. In interaction two, the uncertainties are the opposite. Wages increase, and only risk one occurs. As a result, uncertainties and risks could generate thousands of possible scenarios.

FIGURE 5.7 Probabilistic cost estimation sequence due to uncertainties and risks – Example 2

TABLE 5.4
Risks and Their Probability

Risk	Type	Probability (%)
1- New technology	Opportunity	30
2- Strike	Threat	50
3- Work permission delay	Threat	75

TABLE 5.5
Impacts Inputs for the Triangular Distribution

	Impact Inputs		
	New technology	**Strike**	**Work permission delay**
Least (optimistic)	(11,000)	20,000	9,000
Most likely	(8,000)	25,000	13,000
Most (pessimistic)	(4,000)	25,000	15,000

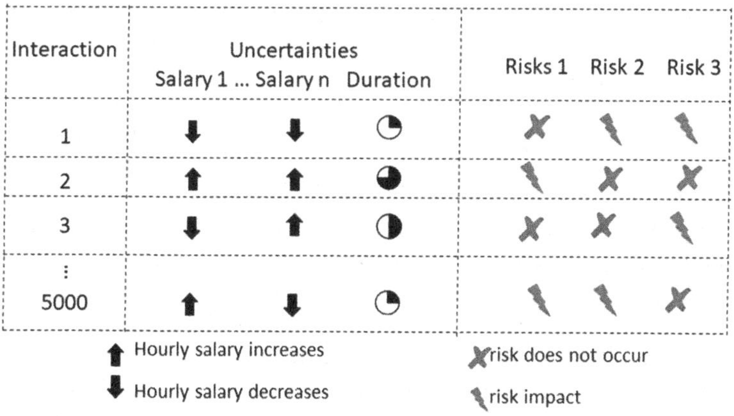

Interaction	Uncertainties Salary 1 ... Salary n Duration			Risks 1	Risk 2	Risk 3
1	↓	↓	◔	✗	⚡	⚡
2	↑	↑	◑	⚡	✗	✗
3	↓	↑	◐	✗	✗	⚡
⋮ 5000	↑	↓	◔	⚡	⚡	✗

↑ Hourly salary increases   ✗ risk does not occur

↓ Hourly salary decreases   ⚡ risk impact

FIGURE 5.8 Possible results for each interaction. It is not exhaustive.

TABLE 5.6
Statistics Result – Example 2

Statistics	Result
Maximum ($)	171,097
Minimum ($)	56,596
Average ($)	111,708
Median ($)	111,701
Percentiles	
10	84,996
Deterministic value ($) P12	88,560
25	96,886
50	110,950
75	124,713
90	137,381

The statistics result is shown in Table 5.6. The range of possible values is from $ 56,596 minimum to $ 171,097 maximum showing that dispersion increased when risks and uncertainties are considered. Also, the average and median are above $ 110,000, highlighting that cost can increase considerably. Considering the inputs adopted in this example, the final statistics could be different because it is a probabilistic calculation, not deterministic (exact). The probabilistic approach does not allow that the same input will have identical results.

Figure 5.9 shows solid and dotted columns. The dotted columns are frequency related to risks and uncertainties, and the solid ones is only for uncertainties from Example 1. Risks promote a right-skewed distribution. The bottom-up cost estimate covers only 12% of the results, as illustrated in the Figure. Hence, the cost overrun is probable and should be mitigated by a contingency discussed in the next section.

5.3   CONTINGENCY

Contingency is a provisioned amount or a time buffer for a project to cover any possible risks that may affect the cost and/or schedule. It is included in the project's baseline, which is calculated and added to the estimated time and cost. Usually, contingency tends to decrease with increasing maturity of the project, and it is expected to be consumed throughout the execution. As proposed by PMI, its use can be controlled by the project manager, and/or by higher levels of the organization, according to GAO.

Table 5.7 shows what is expected to be covered or not by cost contingency.

As each project has specific characteristics, organizations must create procedures to bind what can be considered covered by contingency. By the Table, there are no clear limits established. For example, what would be a slight price fluctuation? It could be 5% or 10%. Consequently, it should be defined by the procedure.

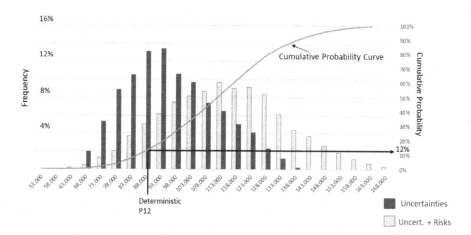

FIGURE 5.9 Probabilistic cost estimate due to uncertainties and risks – histogram of frequencies due to uncertainties, columns, histogram of frequencies due to uncertainties and risks, dotted columns, and cumulative probability curve, solid curve – Example 2

TABLE 5.7

Items Expected or Not to Be Covered by Contingency

Not included on the contingency	Included on contingency
Major changes in contracting strategy.	Changes in the scope and development of engineering design or detailed.
Extraordinary events.	Changes in market conditions.
Major changes in the planning.	Errors and omissions in the project planning or cost estimate.
Escalation and exchange effects.	Variations in environmental conditions.
Exceptional Stand-by.	Small price fluctuations.
Major changes in scope.	Risks associated with the selected technology.
Management reserve and allowances.	

Allowance and contingency are different concepts. The first difference is that allowance is a percentage calculated and applied while preparing the project's cost and schedule estimates. This percentage refers to a part of the scope that is known but was not measured because there weren't sufficient details, or its quantification may not be economically efficient. The contingency is aggregated after the preparation of cost and schedule estimates and is linked to uncertainties and project risks. The allowance will be consumed during the execution of the project, while the contingency may be partially used. Allowance is used only on detailed cost and schedule estimates, not applicable to conceptual estimates (e.g., parametric estimates). On the other hand, contingency can be applied to detailed or conceptual estimates. Figure 5.10 summarizes these differences.

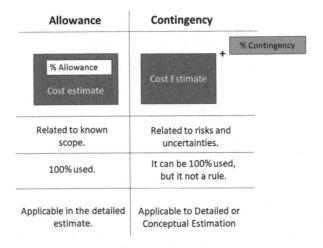

Allowance	Contingency
Related to known scope.	Related to risks and uncertainties.
100% used.	It can be 100% used, but it not a rule.
Applicable in the detailed estimate.	Applicable to Detailed or Conceptual Estimation

FIGURE 5.10   Differences between allowance and contingency

The literature usually finds four methods for cost and time contingency estimates. These methods have advantages and disadvantages according to the project maturity level. In addition, there is the possibility of applying hybrid methods that combine two or more techniques to increase the results' reliability. They are listed below:

- Predetermined percentage
- Expert judgment
- Parametric modelling
- Risk analysis

Predetermined percentage method consists of adopting a predetermined contingency percentage to be added to the cost and schedule estimate for all projects or each type. Typically, the percentage number is recorded in the corporate procedures of each company and can be obtained from predetermined tables. For instance, a contingency percentage defined as 10% in a R$ 10 million project results in a cost contingency of R$ 1 million.

Advantages: Easy application can be used during the entire project phases.

Disadvantages: It can't capture specific project risks. Since this method doesn't relate risk events and contingency amounts, it may be difficult to control the contingency cost use. Although this method applies to all project phases, it is recommended to apply it in early project phases.

Figure 5.11 shows an example. According to the complexity and cost estimate, a fixed percentage is adopted. For example, for high complexity and cost estimate, adopt 30%, and for low complexity and low cost estimate, 10%.

Expert judgement: From an expert's opinion based on his previous experience in project management and risk analysis, it is defined as a contingency

Complexity

FIGURE 5.11   Predetermined percentage example, the percentages are hypothetical.

percentage or amount for time and cost contingency. Obtaining views from multiple experts is recommended to avoid inconsistency and minimize bias.

Advantages: Easy application can be used during all project phases and is commonly used in hybrid combination with other methods.

Disadvantages: Applying the method requires qualified and experienced professionals. It may not capture all specific risks of the project. The expert judgement can be inconsistent or biased.

Parametric model: Based on a project information database, it is possible to generate a parametric model to calculate cost and time contingencies. With technical data, such as production capacity, unit type, and costs, it is possible to establish a correlation among variables and define appropriate contingency for the project. The method uses multivariable regression analysis to determine the equation's correlation coefficients. The simple regression analysis was discussed in Section 3.5.

For example:

Outcome = Constant + coefficient 1*(parameter A) + coefficient 2*(parameter B) + ...

The outcome can be an amount for cost contingency, the parameters represent project data (e.g. technology, production capacity, etc.), and the equation may be linear, logarithmic, exponential, power, or polynomial.

Advantages: Easy application (after defining the parametric equation) and more precise than the two previous methods presented. The technique considers project systemic risks, for example, risks that are not unique to a particular project. These are risks connected to, e.g., company culture, technology adopted, and complexity, among others.

Disadvantages: The method requires a reliable project database. The model definition requires a considerable amount of time. The model may vary depending on the project type. It uses necessarily qualified professionals for the model definition. Although this method applies to all project phases, it is recommended to use it in early project phases. It may not capture all specific risks of the project.

Risk analysis: It is a detailed method and requires experts' opinions and observations to qualify risks and opportunities and to define impact and occurrence probability.

Advantages: The method identifies project-specific risks, correlates risk events and contingency, is more accurate than previous methods, and provides information necessary to increase contingency control levels.

Disadvantages: It is most applicable for a detailed project phase. It requires qualified and experienced professionals for the analysis execution. It is necessary to involve all project team members.

The method may be applied to define integrated cost-time contingency. It's a recommended method because it includes the impact of schedule risk on cost risk and provides more realistic results.

Figure 5.12 summarizes the methods and their characteristics. The time to apply each technique is short for the expert judgment and predetermined percentage but increases considerably for the parametric equation and risk analysis.

The third column represents when the method captures specific risks. Hence, risk analysis is the best method to capture specific and systemic risks. The next column shows what methodology requires training or not, and if the professional needs experience before applying it. Finally, risk analysis and the parametric equation are considered more accurate than expert judgment and predetermined percentage.

5.3.1 CONTINGENCY DUE TO UNCERTAINTIES – EXAMPLE 3

The Example 1 shows Company ABC estimating the probabilistic cost estimate for piping assembly. Suppose that the company wants to define the contingency to ensure that 50% of the simulation has enough money to execute the scope. The P50 result is \$ 92,088 from Table 5.3, and the contingency is the difference between the deterministic value and P50:

$$\text{Contingency} = 92,088 - 88,560 = \$ 3,528$$

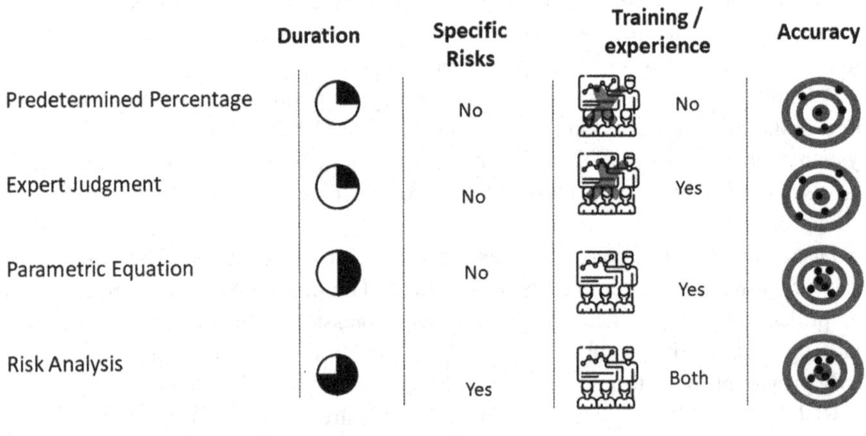

	Duration	Specific Risks	Training / experience	Accuracy
Predetermined Percentage		No	No	
Expert Judgment		No	Yes	
Parametric Equation		No	Yes	
Risk Analysis		Yes	Both	

FIGURE 5.12 Methods and their characteristics and training icon made by Freepik from www.flaticon.com.

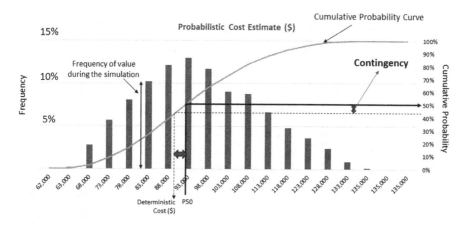

FIGURE 5.13 Contingency adopted to ensure that 50% of simulations in example 1 have enough money to avoid a cost overrun – Example 3

Hence, the contingency estimated is $ 3,528 or 4%, which should be added to the deterministic value. The difference between the two points, P50 and deterministic point, in the cumulative probability curve in Figure 5.13 represents the money needed to conclude the project. This amount represents the contingency.

A different strategy could be adopted if company ABC has a conservative approach. Therefore, they could define the P90 to estimate the contingency. Through Table 5.4, P90 is $ 114,588, meaning that in 90% of interactions, the company is protected from a cost overrun due the uncertainties.

$$\text{Contingency} = 114,588 - 88,560 = \$ 26,028$$

As a result, the contingency should be 29% or $ 26,028, which is a considerable increment.

5.3.2 Contingency Due to Uncertainties and Risks – Example 4

Example 2 shows that company ABC re-estimates the piping assembly through the probabilistic method due to uncertainties and risks. The result shows that only 12% of interactions have enough resources to execute the scope. Hence, a contingency could be added to mitigate the cost overrun threat.

The difference between the deterministic cost estimate and the selected probability in the cumulative probability curve estimates the contingency.

If company ABC adopts the P50 to estimate the contingency, as illustrated in Figure 5.14, the result is:

$$\text{Contingency} = \text{P50} - \text{P12 (deterministic cost)}$$
$$\text{Contingency} = 111,701 - 88,560 = 23,141$$

The result is that the contingency should be 26% or $ 23,141. It confirms that risks increased the contingency amount significantly. Also, as mentioned before, the

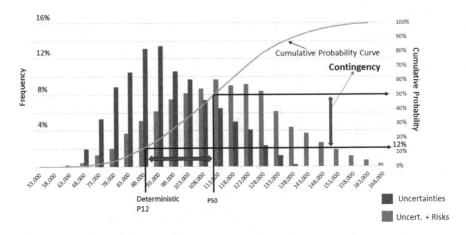

FIGURE 5.14 Contingency adopted to ensure that 50% of simulations in example 2 have enough money to avoid a cost overrun – Example 4

TABLE 5.8
Impacts Inputs for the Triangular Distribution – Mitigated Scenario

	Impact Inputs – Reassessment		
	New technology	**Strike**	**Work permission delay**
Least (optimistic)	(11,000)	15,000	7,000
Most likely	(8,000)	20,000	10,000
Most (pessimistic)	(4,000)	22,000	12,000

---

manager's decision of which point of the cumulative distribution curve should be adopted is related to market and business conditions.

In addition, it is possible to create a mitigated scenario where a mitigation action is implemented for each risk. It allows a new impact and likelihood assessment for each risk and, finally, a new simulation, whereas the contingency amount could be reduced.

Then, Company ABC defines actions and plans to mitigate the strike and work permission risks, permitting a new assessment.

Strike likelihood reduces from 50% to 30%, and work permission delay from 75% to 50%. And Table 5.8 shows the cost impact reviewed.

After the inputs are reviewed, a new simulation is generated, and the result is illustrated in Figure 5.15 and statistics are shown in Table 5.9.

It shows that P50 has improved considerably, and the new contingency is calculated below:

$$\text{Contingency} = \text{P50} - \text{P25 (deterministic cost)}$$
$$\text{Contingency} = 100{,}380 - 88{,}560 = \$\ 11{,}820$$

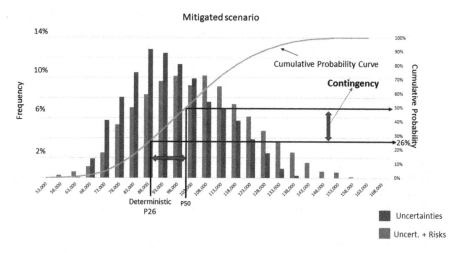

FIGURE 5.15 Contingency – Mitigated Scenario – Example 4

TABLE 5.9
Statistics Result – Mitigated Scenario – Example 4

Statistics	Result
Maximum ($)	162,515
Minimum ($)	54,681
Average ($)	101,569
Median ($)	100,380
Percentiles	
10	78,185
25	88,088
Deterministic value ($) P26	88,560
50	100,380
75	114,669
90	126,443

The mitigated scenario reduced the contingency from 26% to 14% by adopting the P50.

5.4 MANAGEMENT RESERVE

Although management reserve (MR) has similarities with contingency because both aim to avoid cost overrun, they do not have the same meaning. According to AACE International, management reserve is defined as "(…) *amounts that are within the defined scope, but for which management does not want to fund as contingency or that cannot be effectively managed using contingency.*"

Management reserve is related to portfolio. Consequently, it is managed by the organization to protect its projects. Contingency must be included in the project baseline, but MR is not included in the baseline of each project. It is used if the project contingency is insufficient to execute the scope because one or more portfolio risks are occuring. Figure 5.16 summarizes the differences between the management reserve and contingency.

Figure 5.17 exemplifies the management reserve. Suppose Company ABC has three projects in their portfolio: A, B, and C. Also, each project has its specific contingency, contingency A, B, and C, respectively. The management reserve is estimated to protect all projects, represented by a curly bracket. Despite that, the assumption is that only B requests the management reserve, rectangle, which means that project B's estimate and the contingency were insufficient.

The method to estimate the management reserve is divided into four stages, like in Figure 5.18.

Portfolio calculation is based on the defined total amount the company is expected to spend on all projects in a specific period, the fiscal year, for example. Each project budget should be added to the respective contingency.

	Management Reserve	Contingency
Project Baseline	Not Included → Management Reserve / Included	% Contingency
	Used in Portfolio Management	Used in Project management
	Management is done by the organization	Management is done by the Project Manager

FIGURE 5.16   Differences between management reserve and contingency

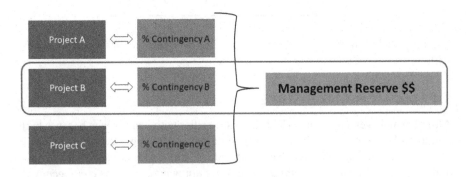

FIGURE 5.17   Management reserve example

| Portfolio Calculation | Portfolio Risk Identification | Portfolio Risk Analysis | Management Reserve Estimation |

FIGURE 5.18 Sequence of management reserve estimation

TABLE 5.10
Company Blue Sea Portfolio

Project's name	2023 Budget ($)	Contingency (%)	Annual budget + contingency
Sea	4,400,000	20 or 880,000	5,280,000
Dolphin	3,350,000	10 or 335,000	3,685,000
Whale	1,750,000	10 or 175,000	1,925,000
Big Island	4,200,000	20 or 840,000	5,040,000
Portfolio forecast	13,700,000		15,930,000

Portfolio risk identification is based on identifying the opportunities and threats, but one attention point is to avoid duplication. It means that risks identified in each project risk identification should not be considered in this stage. Examples of portfolio risks include force majeure, subsidies, and high inflation.

The risk analysis defines the likelihood and impact of each risk. After that, these inputs are included in the cost model to run the Monte Carlo simulation, as illustrated in Example 3.

In the end, the management reserve is estimated by Equation 5.1.

$$MR = \sum Px - \sum PF \qquad \text{(Equation 5.1)}$$

Where,
MR = Management reserve
Px = Probability adopted
PF = Portfolio forecast

5.4.1 MANAGEMENT RESERVE ESTIMATION – EXAMPLE 5

Company Blue Sea wants to estimate the management reserve for 2023. They have four projects in their portfolio, as listed in Table 5.10.

The portfolio forecast is $ 15,930,000 in 2023, and five risks are identified in Table 5.11.

The opportunities are tax-free for some activities related to the company portfolio, and the second is not a "real" opportunity. It relates to one or more projects, having low performance and spending less than planned. Consequently, it will promote a negative impact and "save" money.

Threats are force majeure or extreme events, such as hurricanes or floods. Also, a change in economic premises could be a high escalation due to war, for example.

TABLE 5.11
Opportunities and Threats

Risk	Type	Probability (%)
1- Tax-free	Opportunity	30
2- Low performance	"Opportunity"	75
3- Change in economic premises	Threat	50
4- Force majeure	Threat	10
5- Contingency does not enough	Threat	25

TABLE 5.12
Impacts inputs for Triangular Distribution

	Impact Inputs				
	1-Tax-free	2-Low performance	3- Change in economic premises	4-Force majeure	5-Contingency not enough for one or more projects
Lowest (optimistic)	(1,500,000)	(800,000)	1,000,000	2,000,000	200,000
Most likely	(1,000,000)	(800,000)	3,000,000	4,000,000	500,000
Highest (pessimistic)	(500,000)	(200,000)	5,000,000	6,000,000	800,000

Finally, the opposite of low performance is when the project has a cost overrun and the contingency is insufficient. The company should use the management reserve to cover it in this case.

Each of these risks should be defined as an occurring probability. For example, tax-free has a 30% of incidence, meaning that for every 100 simulations, the risk will impact the portfolio in 30 cases.

Then the impact is defined through the triangular distribution, as listed in Table 5.12. The opportunities have a negative input because they reduce the spending in the year. As mentioned before, risks should not be included in this analysis if they have already been included in the project risk analysis, because it will generate a duplication or redundancy.

Table 5.13 extracts 20 probabilistic results to explain how it works. Each interaction shows when the risk occurs, meaning a number different from zero is a risk occurrence. In these cases, the impact is between the range defined in the triangular distribution, as shown in Table 5.12.

The total column is the portfolio forecast sum plus each risk's impact. Hence, interactions 16 and 20 show cases where no risks occur, resulting in the portfolio forecast of $ 15,930,000.

The Monte Carlo simulation result is plotted in Figure 5.19 after 5,000 interactions. Columns are the frequency histogram showing each frequency along the

TABLE 5.13
Sample of 20 Results of 5,000 Simulations

ID	Portfolio forecast ($ thousand)	1-Tax-free ($ thousand)	2-Low performance ($ thousand)	3-Change in economic premises ($ thousand)	4-Force majeure ($ thousand)	5-Contingency not enough for one or more projects ($ thousand)	Total ($ thousand)
1	15,930	-	-	2,305	-	-	18,235
2	15,930	(1,306)	-	4,224	-	-	18,848
3	15,930	-	(483)	2,090	-	-	17,537
4	15,930	-	(598)	2,531	-	-	17,863
5	15,930	-	(802)	3,378	-	-	18,506
6	15,930	-	(770)	3,216	-	-	18,376
7	15,930	-	(698)	-	-	-	15,232
8	15,930	-	(624)	2,631	-	-	17,938
9	15,930	-	(504)	-	-	375,450	15,802
10	15,930	(1,245)	-	-	-	-	14,685
11	15,930	-	(758)	3,161	-	-	18,333
12	15,930	-	-	-	3,701	-	19,631
13	15,930	-	(701)	2,928	-	-	18,157
14	15,930	-	(536)	-	-	-	15,394
15	15,930	-	(400)	-	-	-	15,530
16	15,930	-	-	-	-	-	15,930
17	15,930	(1,170)	-	-	-	-	14,760
18	15,930	(1,211)	-	3,844	-	-	18,563
19	15,930	-	(683)	2,858	3,858	-	21,964
20	15,930	-	-	-	-	-	15,930

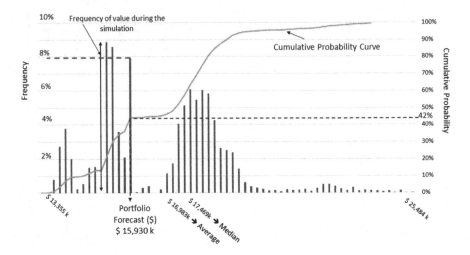

FIGURE 5.19 Portfolio probabilistic results of company blue sea – histogram of frequencies, columns, and cumulative probability curve, solid curve.

TABLE 5.14
Statistical Results

Statistics	Result
Maximum ($)	25,484,624
Minimum ($)	13,355,266
Average ($)	16,983,866
Median ($)	17,469,029
Portfolio Forecast ($)	15,930,000

simulation. For example, the portfolio forecast, $ 15,930, has around 8% of frequency, as shown in the dashed line. It means that about 400 interactions have the exact portfolio forecast result. Also, the range of possible values is between $ 13,355 (minimum) and $ 25,484 (maximum).

Furthermore, there is the cumulative probability curve, the solid curve. The dashed line shows that portfolio forecast (PF) performs 42% of the simulations. It means that Company Blue Sea has enough money to execute its projects in 2023 for 42% of possible scenarios.

Table 5.14 shows the results of the maximum, minimum, average, and median statistics. The range of possible values is from $ 13.4 million to $ 25.5 million. It highlights that the median, $ 17.5 million, and the average of $ 17.0 million is over the PF, which shows that a positive management reserve is recommended.

The management reserve estimation is done by Equation 5.1. Company Blue Sea should choose the P(x) from the cumulative probability curve considering the risk appetite. The assumption is that the company wants to ensure that the management reserve covers half of the simulations (or P50).

$$MR = P50 - portfolio\ forecast$$

$$MR = 17,469k - 15,930$$

$$MR = \$\ 1,539,000$$

Hence, Company Blue Sea needs $ 1.5 million (10%) as a management reserve for 2023. Figure 5.20 shows the management reserve necessary to ensure that 50% of interactions have enough budget to execute.

Also, if the company Blue Sea has a high-risk appetite, they can decide on a low point in the cumulative probability curve, P30, for example. Through Equation 5.1, the management reserve is:

$$P30 = \$\ 15,461$$

$$MR = P30 - portfolio\ forecast$$

$$MR = 15,461 - 15,930$$

$$MR = \$ - 469,000$$

Figure 5.21 shows that the company Blue Sea will not need the total amount calculated by the portfolio forecast because they believe that the opportunities mapped will be explored and the threats will be mitigated.

Another option is to adopt a conservative approach, which could be corroborated by the right skew of the histogram illustrated in Figure 5.22. The Government Accountability Office, US, 2020, mentions that P65 or a higher point in the cumulative probability curve is recommended. Through the P65, the management reserve is $ 2 million, as below.

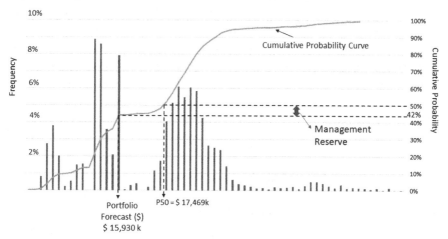

FIGURE 5.20   Management reserve estimation

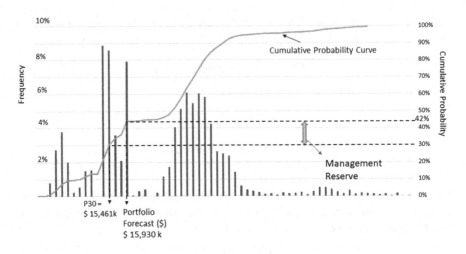

FIGURE 5.21   Management reserve estimation – High risk appetite- P30 Adopted.

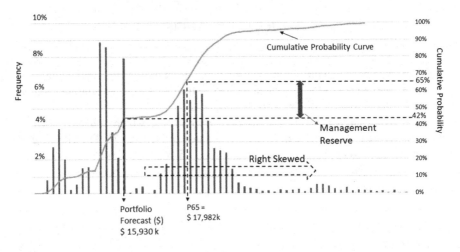

FIGURE 5.22   Management reserve estimation – conservative approach related with right skewed.

$$P65 = \$ \ 17{,}982$$

$$MR = P65 - portfolio \ forecast$$

$$MR = 17{,}982 - 15{,}930$$

$$MR = \$ \ 2{,}052{,}000$$

Consequently, the selected point on the cumulative probability curve is ultimately a manager's decision related to business and market conditions.

BIBLIOGRAPHY

AACE International, 2008. *Recommended Practice 40RS-08 Contingency Estimating – General Principles.* Available at: https://www.pathlms.com/aace/courses/2928. [Accessed: 20 March 2023].

AACE International, 2008. *Recommended Practice 57R-09, 2011, Integrated Cost and Risk Analysis using Monte Carlo Simulation of a CPM Model.* Available at: https://www.pathlms.com/aace/courses/2928. [Accessed: 20 March 2023].

AACE International, 2022. *Recommended Practice 10S-90 Cost Engineering Terminology.* Available at: https://web.aacei.org/resources/cost-engineering-terminology. [Accessed: 8 January 2023].

GAO, 2020. Cost Estimating and Assessment Guide. Available at: https://www.gao.gov/products/gao-20-195g. [Accessed: 15 December 2022].

Hastak, M., 2015. *Skills & Knowledge of Cost Engineering,* 6th Edition. AACE International.

PMI – Project Management Institute, 2021. *PMBOK Guide. A Guide to the Project Management Body of Knowledge,* 7th Edition. Project Management Institute, Inc.

6 Economic Costs

The economic indicators aim to show the project's benefits and make comparisons. In addition, they can support selecting the best solutions aligned with the strategic company goals.

They should be used in the planning phase to support the decision gates. Still, they can be used in the execution phase, analyzing improvements or supporting the decision process, as a possible plant shutdown from the break-even analysis.

The attention point is input quality data because the result is not applicable if the data is unreliable. Also, the estimation should be registered and traceable for future audits and as historical information.

The chapter shows a group of indicators listed below, definitions, formulas, advantages and disadvantages, examples, and exercises.

- Payback period
- Return on investment (ROI)
- Return on capital employed (ROCE)
- Break-even analysis
- Net present value (NPV)
- Profitability index (PI)

6.1 PAYBACK PERIOD

The payback period defines the time to recover the investment cost of a project or asset. Thus, the project should aim for the shortest period. It is important to alert that it is a straightforward method and can prove inaccurate.

It is estimated by Equation 6.1:

$$\text{Payback period} = \frac{\text{Investment Cost}}{\text{Gains per year}} \qquad \text{(Equation 6.1)}$$

Investment cost is the original amount of money to implement the project, such as a new road or airport.

Gains are the profit after direct and indirect costs are deduced from the revenue for a given period (e.g., 1 year).

Table 6.1 lists the main advantages and disadvantages.

Time value of money (TVM) means that today's money has a higher value than in the future because of the money's earning potential.

6.1.1 WATER TREATMENT PLANT – EXAMPLE 1

The example is based on two options for a water treatment plant: open and closed circuits, like in Figure 6.1. An open circuit means that the water is collected, then

DOI: 10.1201/9781003402725-6

TABLE 6.1

Advantages and Disadvantages of the Payback Period Method

Advantages	Disadvantages
Quick method	Ignores time value of money (TVM)
Few inputs	Not all cash flows are covered
Preference to liquidity	A project can have a reasonable payback period and a negative NPV

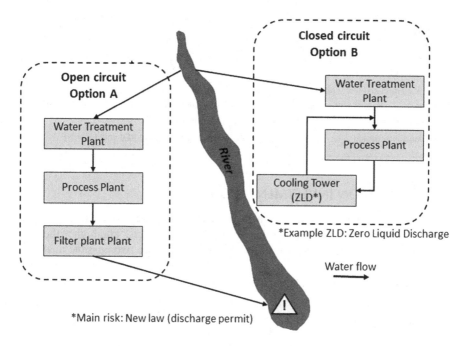

FIGURE 6.1   Open and closed circuit for a water treatment plant

used in the process plant, and finally, it is discharged into the sea or river. The closed circuit ensures that the water is not discharged, reducing the risk of pollution in the river or sea.

The data and the estimation is shown in Table 6.2.

The open circuit A shows a better result, but the decision should be made considering a group of indicators and risks. For example, the open circuit has a new risk of changing the discharge permit. The case is reviewed in Section 6.2.

6.1.2   PAYBACK PERIOD – EXERCISE 1

Estimate the payback period to a new process plant, in which the investment is $ 100 million and gains per year are $ 20 million.

TABLE 6.2
Payback Period Example

	Open circuit A	Closed circuit B
Investment cost	$ 200 million	$ 250 million
Gains per year	$ 50 million	$ 50 million
Payback period	200/50 = 4 years	250/50 = 5 years

TABLE 6.3
Advantages and Disadvantages of ROI

Advantages	Disadvantages
Quick method and few inputs	Ignores time value of money (TVM)
A simple measure that allows project comparisons	To compare companies, they should use the exact definition for gains and investment costs.
Universally accepted	A project can have a reasonable ROI and a low NPV

6.2   RETURN ON INVESTMENT (ROI)

Return on investment is a ratio or a metric that shows the investment's gains. It can be used to draw a comparison between different projects. The higher the ROI percentage, the higher the return to the company. There are two possible analyses: simple and complex ROI.

Both equations are shown below:

$$\textbf{Simple ROI} = \frac{\text{Gains} \times \text{Operation Life} - \text{Investment Costs}}{\text{Investment Costs}} \qquad \text{(Equation 6.2)}$$

$$\textbf{Complex ROI} = \frac{\text{Average yearly profit during earning life}}{\text{Investment Costs} + \text{working capital}} \qquad \text{(Equation 6.3)}$$

Where:

Investment cost is the original amount of money needed to implement the project, as with a new road or airport.

Gains is the profit after direct and indirect costs are deduced from the revenue for a given period (e.g., 1 year).

Working capital is the difference between the assets and liabilities. It means that the current cash, goods, raw material, and invoices to be received are the asset and current debts and wages are liabilities.

Table 6.3 lists the main advantages and disadvantages.

6.2.1 WATER TREATMENT PLANT – ROI – EXAMPLE 2

Estimate the simple ROI for two options of water treatment industrial plant, like Figure 6.2. The difference is based on the filter technology adopted.

Option A uses better technology for the filtration stage, and option B uses a traditional and cheaper filter.

The data and the estimation is shown in Table 6.4.

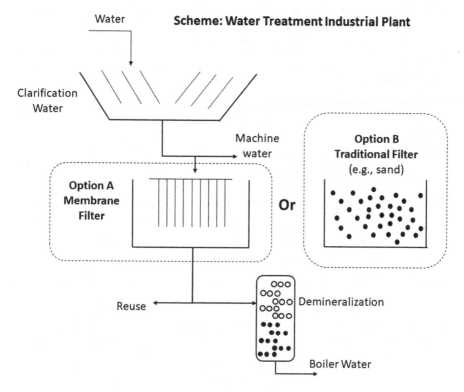

FIGURE 6.2   Water treatment plant options: a – membrane filter, b – traditional sand filter

TABLE 6.4
ROI Example 1

	Unit	Option A	Option B
Operation	Years	10	10
Gains	$ million per year	15	10
Investment costs	$ million	100	80
Simple ROI	%	$= \dfrac{15 \times 10 - 100}{100} = 50\%$	$= \dfrac{10 \times 10 - 80}{80} = 25\%$

The results show that option A has a higher investment than option B, but the ROI of opportunity A is double that of option B, meaning it is a better choice.

6.2.2 Scenario Analysis – ROI – Example 3

Estimate the simple ROI for the water treatment plant proposed in Section 6.1.1. Also, a second scenario should be created for the open circuit considering that the risk of discharged permit occurs, as illustrated in Figure 6.1, considering the cost impact of the risk is US$ 12 million per year.

The data is shown in Table 6.5:

TABLE 6.5
ROI Example 3

	Unit	Open circuit (risk NOT occurs)	Open circuit (risk occurs)	Closed circuit
Operation	Years	10	10	10
Gains	$ million per year	50	50	50
Risk impact	$ million per year	0	–12	Not applicable
Investment costs	$ million	200	200	250
Simple ROI	%	$= \dfrac{50 \times 10 - 200}{200}$ $= 150\%$	$= \dfrac{(50\text{-}12) \times 10 - 200}{200}$ $= 90\%$	$= \dfrac{50 \times 10 - 250}{250}$ $= 100\%$

Working capital: US$ 20 million
CAPEX: US$ 330 million

The result shows that the open circuit has a better ROI, 150%, than the closed circuit, 100%, when the risk is not considered. However, the cost impact affects the ROI if the risk occurs, reducing it to 90%. It shows that risk should be considered, and a risk analysis should be done before any decision.

6.2.3 Coal and Solar – ROI – Exercise 2

Scenario A: Two power plants of 10 MW using different sources should be analyzed considering the type below: coal and solar. Choose the best option through the data from Table 6.6.

Scenario B: Repeat the process considering that coal subsidies are removed, which means a 10% reduction in the gains per year. Also, the power plant receives new subsidies, which means a 10% increment in the gains per year.

6.2.4 Complex ROI – Exercise 3

The Sky Blue company has a performance, like Table 6.7, in the last 10 years. What is the complex ROI considering the CAPEX and working capital below?

TABLE 6.6
ROI Exercise 2

	Unit	Coal power plant	Solar power plant
Operation	Years	25	25
Gains	$ million per year	4	3.5
Investment costs	$ million	35	37

TABLE 6.7
Data ROI Exercise 3

Year	Profit (US$ million)
1	20
2	30
3	50
4	60
5	80
6	80
7	60
8	50
9	50
10	40

6.3 RETURN ON CAPITAL EMPLOYED (ROCE)

Return on capital employed (ROCE) is a financial ratio between the EBIT and capital employed. It measures the profit of a company as a percentage of the capital.

It is a helpful indicator to compare companies in the same industry. Also, it should be done in the same year.

The ROCE equation is shown below.

$$\text{ROCE} = \frac{\text{EBIT}}{\text{Capital Employed}} \qquad \text{(Equation 6.4)}$$

Where:
  EBIT = Earnings before interest and tax
  EBIT = Revenue – operating expenses – cost of goods
  Capital employed = Total assets – current liabilities

6.3.1 ROCE – Example 4

Table 6.8 shows that company Green has a small revenue than Company Blue. Although Company Blue is much bigger than Company Green, the ROCE from Company Green, 33%, is much higher than Company Blue, 14%. It shows that each dollar applied in Company Green has a better return.

Also, in comparison with the industry benchmark, as shown in Figure 6.3, both companies have a worse result because their ROCE is below the benchmark, 38%.

6.3.2 ROCE – Exercise 4

Company Sky Blue is looking to acquire a new company, and there are four options: ST Plus, ZKV Company, YES Tecom, and ALL Company. One of the targets is a ROCE above 30%. What are the options that match the goal? Table 6.9 shows the data for the last year.

6.4 BREAK-EVEN ANALYSIS

Break-even analysis is important to define when the shutdown must occur at the process plant. "*In business operations, the Break-Even point is the rate of operations output or sales at which income is sufficient to equal operating costs or operating cost plus additional obligations that may be specified*" (AACE International, 2022).[1]

The Equation 6.5 defines the break-even point.

$$BP = FC + VC = TR \qquad \text{(Equation 6.5)}$$

Where
BP = Break-even point
FC = Fixed cost, as rent, depreciation and insurance.

TABLE 6.8
ROCE Estimation – Example 4

Company	Unit	Blue	Green
Revenue	$ million	6	1.5
Operating expenses	$ million	2.5	0.6
Cost of goods	$ million	2.5	0.5
EBIT	**$ million**	6 – 2.5 – 2.5 = 1	1.5 – 0.6 – 0.5 = 0.4
Total assets	$ million	10	2
Current liabilities	$ million	3	0.8
Capital employed	**$ million**	10 – 3 = 7	2 – 0.8 = 1.2
ROCE	%	1/7 = 14%	0.4/1.2 = 33%

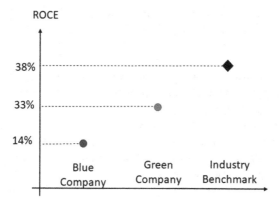

FIGURE 6.3 ROCE comparison through industry benchmark – Example 4

TABLE 6.9
ROCE Data – Exercise 4

ST Plus

Revenue	50	US$ million
Operating expenses	17	US$ million
Cost of goods	25	US$ million
Total assets	35	US$ million
Current liabilities	10	US$ million

ZKV Company

Revenue	70	US$ million
Operating expenses	30	US$ million
Cost of goods	35	US$ million
Total assets	45	US$ million
Current liabilities	25	US$ million

YES Tecom

Revenue	60	US$ million
Operating expenses	20	US$ million
Cost of goods	30	US$ million
Total assets	38	US$ million
Current liabilities	10	US$ million

All Company

Revenue	90	US$ million
Operating expenses	40	US$ million
Cost of goods	45	US$ million
Total assets	45	US$ million
Current liabilities	20	US$ million

VC = Variable cost, as raw material. It means that cost depends on daily production; if production increases, the variable cost also rises.

TR = Total revenue

Figure 6.4 summarizes the break-even analysis. If the result of Equation 6.5 is lower or equal to the break-even point, the plant should shut down because the operation is not economically feasible.

The break-even analysis can improve profitability by reducing fixed and variable costs and raising revenue. The method has disadvantages because it considers that production equals sales, which could not be an actual situation. Also, it can be complex to mix different products generated by the factory.

6.4.1 Break-even Analysis – Example 5

The example is based on an oil platform operation, and the assumption is that Brent price is always US$ 35. What should be the daily production of BOE (barrel of oil equivalent) to achieve a zero break-even?

The fixed cost (FC) to support the operation is US$ 2,500,000, and the variable cost (VC) is calculated by the equation below:

$$VC = DP \times 10$$

Where, DP = Daily production

The revenue is estimated by Brent price multiplied by daily production. Considering this information, the break-even production can be calculated.

The break-even point means that the total revenue is the equal total cost. Hence,

$$\text{Total revenue} = \text{Brent} \times \text{daily production} = 35 \times DP$$

$$\text{Total costs} = FC + VC = 2,500,000 + DP \times 10$$

In the break-even point:

$$35 \times DP = 2,500,000 + DP \times 10 \to 35\ DP - 10\ DP = 2,500,000 \to 25\ DP = 2,500,000$$

$$\textbf{DP = 100,000 BOE}$$

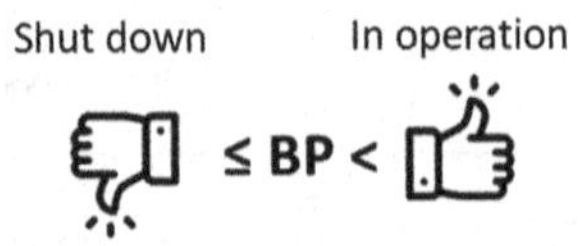

Shut down   In operation

≤ BP <

FIGURE 6.4   Break-even analysis and icons made by Freepick from www.flaticon.com

The plant should be shut down if the daily production is below 100,000 BOE. If it is above, the plant can operate because it is profitable. Figure 6.5 summarizes it. The break-even point is the intersection between the revenue and total cost curve.

6.4.2   Brent Analysis – Example 6

If the oil platform has a daily production of 50,000 BOE, what should be the Brent price to achieve a zero break-even?

The fixed cost (FC) to support the operation is US$ 2,000,000, and the variable cost (VC) is calculated by the equation below:

$$VC = \text{Daily production} \times 8$$

Hence, the VC is fixed because the daily production is constant. The revenue is estimated by Brent price multiplied by daily production (DC). The Brent price varies from US$ 20 to US$ 60.

Considering this information, the break-even Brent can be estimated.

The total revenue is equal total cost at the break-even point. Hence,

Total revenue = Brent × daily production
Total cost = FC + VC = 2,000,000 + DP × 8
Total cost = 2,000,000 + 50,000 × 8 = 2,400,000
At the break-even point, total costs = Total revenue
Total revenue = Brent × DP
Brent × 50,000 = 2,400,000

Brent = US$ 48

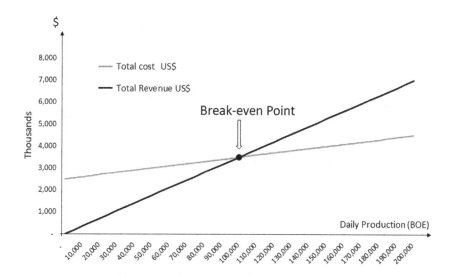

FIGURE 6.5   Total costs and total revenue by daily production

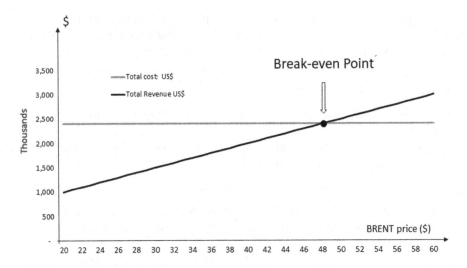

FIGURE 6.6   Total costs and total revenue by Brent Price

Hence, if the Brent price is below US$ 48, the plant should be shut down. If it is above, the plant can operate because it is profitable. Figure 6.6 summarizes it. The break-even point is the intersection between the revenue and total cost curve. The example explains why the Brent price affects the oil industry worldwide.

6.4.3   BREAK-EVEN ANALYSIS – EXERCISE 5

A thermal gas plant generates electricity and has capacity of 10 MW. The electricity price is estimated at US$ 0.20 per kWh. The fixed cost (FC) is US$ 20,000 per day, and the variable cost (VC) is calculated by the equation below:

$$VC = 20,000 + capacity\ (MW) \times 1,000 \times \log gas\ price$$

Considering that the plant operates in the nominal capacity, 10 MW, and the total revenue per day is capacity × 24h × eletricity price, estimate the gas price in the break-even point.

Disclaimer: the variable cost equation is hypothetical. The only purpose is to exercise the break-even point estimation.

6.5   NET PRESENT VALUE (NPV)

The net present value (NPV) is a method which compares the differences between the total present value of all cash inflows and cash outflows minus the CAPEX. If the NPV is positive it means that investment will be feasible. If it is negative, the project should be canceled or reviewed. It is one of the best techniques for analyzing the profitability of an investment.

NPV has the critical advantage that considers the time value of money (TVM), which means the amount of money in the future does not have the same value as today. Today, one dollar applied to the stock market can result in a higher profit than an investment project, like an O&G plant.

NPV allows comparisons using the interest rate. It is defined as *"the ratio of the interest payment to the principal for a given unit of time and is usually expressed as a percentage of the principal"* (AACE International, 2022).[2]

Each spending in the future should be "moved" to the present using one of the equations below:

Single Payment, Equation 6.6:

$$A \times \frac{1}{\left(1+i\right)^{t}}$$
(Equation 6.6)

Uniform series, Equation 6.7:

$$A \times \frac{\left(1+i\right)^{t}-1}{i \times \left(1+i\right)^{t}}$$
(Equation 6.7)

Arithmentic Gradient, Equation 6.8:

$$A \times \left\{ \frac{\left(1+i\right)^{t}-1}{i^{2} \times \left(1+i\right)^{t}} - \frac{t}{i \times \left(1+i\right)^{t}} \right\}$$
(Equation 6.8)

Where:
A = Amount of money. The amount can be positive (e.g., revenue) or negative (e.g., spent as maintenance)
t = Number of time, for example, years.
i = Interest rate per period (%)

The single payment occurs at a unique time like, for example, the decommissioning that happens in a special year, as in Figure 6.7.

Through the Equation 6.6 and considering an interest rate, i%, the spend is moved to the present, like in Figure 6.8. It means that at present, the small arrow is lower than in the future because this amount of money applied through the interest rate, i%, generates in 5 years, the long arrow.

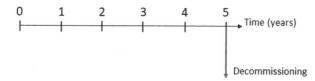

FIGURE 6.7 Single payment

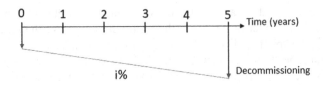

FIGURE 6.8　Single payment occurred in 5 year is moved to 0 year

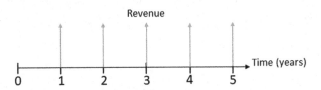

FIGURE 6.9　Uniform series

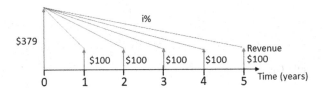

FIGURE 6.10　Uniform series – The revenue is "Moved" to the year 0 at i% rate

The uniform series is useful when the same amount of money is spent each year, like revenue in Figure 6.9. It is possible to see that arrows have the same length.

The example in Figure 6.10 shows that each revenue per year is moved to the present using Equation 6.7 and considering an interest rate, i%. To exemplify the difference, if the revenue is US$ 100 per year, after 5 years, the total revenue is US$ 500. However, using Equation 6.7 and an interest rate of 10%, just US$ 379 is necessary now to generate the same amount of money. As a result, if the company applies $ 379 at a 10% rate, it will result in the same amount of money, $ 500, after 5 years.

$$NPV = 100 \times \frac{(1+0.1)^5 - 1}{0.1 \times (1+0.1)^5}$$

$$= 100 \times 3.79$$

$$= US\$ 379$$

The gradient series consists of the same amount of money increasing (or decreasing) each year, like Figure 6.11. For example, the sales are $ 100 in year 1, $ 200 in year 2, and successively, so in year 5, the sales are $ 500. Also, the total after five years is US$ 1,500.

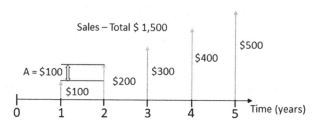

FIGURE 6.11 Gradient series

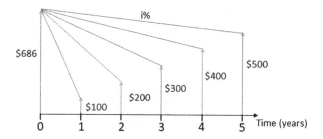

FIGURE 6.12 Gradient series – The sales are "Moved" to the year 0

The example in Figure 6.12 shows that through the Equation 6.8 and considering an interest rate, i% of 10%, each spend per year can be moved to the present. Hence, the NPV is:

$$NPV = 100 \times \left\{ \frac{(1+0.1)^5 - 1}{0.1^2 \times (1+0.1)^5} - \frac{5}{0.1 \times (1+0.1)^5} \right\}$$

$$= \$686$$

Hence, if the company applies $ 686 through 10%, they receive the same amount of money that product sales generate.

6.5.1 NPV – EXAMPLE 07

The Company Green must pay US$ 10 million after 10 years of operation as a unique event. The director asks how much money the company should save now, considering the 5% of interest rate.

The example is a single payment example and should use Equation 6.6. Hence, the NPV is:

$$NPV = 10,000,000 \times \frac{1}{(1+0.05)^{10}}$$

$$= US\$6,139,133$$

Hence, the company must save $ 6.14 million.

6.5.2 SCENARIO ANALYSIS – EXAMPLE 8

Company Blue must choose the best maintenance contract for the new power plant, whose operation life is 25 years. They have two options described below:

Option A – Pay US$ 1,000,000 per year. It is a uniform series example.

Option B – Pay through a gradient series with an increment of US$ 150,000. It means that the first year they pay US$ 150,000, and the second US$ 300,000, and so on.

Considering the 10% of interest rate and through Equations 6.7 and 6.8, we can estimate the NPV for each option:

$$NPV \text{ option } A = 1,000,000 \times \frac{(1+0.1)^{25}-1}{0.1 \times (1+0.1)^{25}}$$

$$= 1,000,000 \times 9.077040$$

$$= US\$ 9,077,040$$

$$NPV \text{ option } B = 150,000 \times \left\{ \frac{(1+0.1)^{25}-1}{0.1^2 \times (1+0.1)^{25}} - \frac{25}{0.1 \times (1+0.1)^{25}} \right\}$$

$$= US\$ 10,154,460$$

Even though option B will pay a low amount of money in the first years, the results show that option A has a lower NPV, and the difference for 25 years is over 1 million.

6.5.3 NPV – EXERCISE 6

Company Sky Plus is seeking a new operation contract for new solar farms. The contract duration is 20 years, and the interest rate is 3%. Two options are available, and the director asks what the best option is through the NPV method.

Data

Option A – A single payment in year 1 of US$ 1 million plus an annual uniform payment of US$ 120,000.

Option B – A gradient payment of US$ 120,000 in year 1, US$ 240,000 in year 2, and so on.

6.6 PROFITABILITY INDEX (PI)

The profitability describes a comparison between total present values of cash inflows and the initial investment.

The formula of PI is according to Equation 6.9:

$$PI = \frac{NPV}{CAPEX} \qquad \text{(Equation 6.9)}$$

Where:

NPV = Net present value

CAPEX = Capital expenditures

The PI helps select projects in a portfolio. It means that the project which has the highest PI should be chosen. However, there is caution because the PI does not differentiate the project scale. For example, two projects with 1 billion or 1 million CAPEX can have the same PI.

Table 6.10 lists the main advantages and disadvantages.

6.6.1   PROJECT SELECTION – EXAMPLE 9

Company Sky Blue is analyzing its portfolio. They have five projects to execute next year, listed in Table 6.11, and their budget is US$ 1.2 million. However, it is not enough to support all projects.

The project selection should be made through PI higher than 0.5.

Hence, projects A, B, and E have a PI higher than 0.5 and must be selected. The CAPEX selected totalized US$ 1,150 below the budget. Figure 6.13 summarizes the result, the brace show three chosen projects, and the dashed circle represent the two projects with low PI and excluded the following year's portfolio.

TABLE 6.10
Advantages and Disadvantages of PI

Advantages	Disadvantages
The calculation is easy to be done.	Two projects of different scales, higher and lower investments, can have the same PI.
Useful indicator to project selection and prioritarization.	Do not cover other project aspects such as quality, safety, etc.
It considers the TVM.	

TABLE 6.11
Data and PI Estimation – Example 9

Project	CAPEX (US$ Thousand)	NPV (US$ Thousand)	PI
Project A	700	500	0.71
Project B	350	180	0.51
Project C	200	50	0.25
Project D	180	60	0.33
Project E	100	60	0.60
Total	1,530	-	-

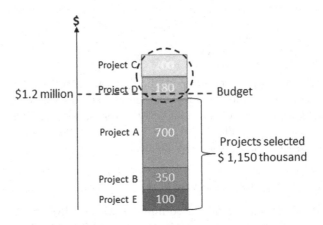

FIGURE 6.13   Project selection – Example 9

TABLE 6.12
Projects List – Exercise 7

Project	CAPEX (US$ thousand)	NPV (US$ thousand)
Lion 01	15	12
Lion 02	50	30
Lion 03	80	30
Mouse 01	100	55
Mouse 02	190	100
Dog 01	210	150
Dog 02	220	100
Cat 01	230	160
Cat 02	240	140
Total	1,335	-

6.6.2   PI – Exercise 07

The company Pets SA must define the following year's projects considering a budget of US$ 800,000. The portfolio has nine projects and a US$ 1,015 planned. Select projects with a PI higher than 0.5 which respect the budget.

NOTES

1. Reprinted with permission from AACE International. Check the website for the latest versions (https://web.aacei.org/resources/cost-engineering-terminology).
2. Reprinted with permission from AACE International. Check the website for the latest versions (https://web.aacei.org/resources/cost-engineering-terminology).

BIBLIOGRAPHY

AACE International, 2022. *Recommended Practice 10S-90 Cost Engineering Terminology.* Available at: https://web.aacei.org/resources/cost-engineering-terminology. [Accessed: 8 January 2023].

Chen, J., 2023. *Profitability Index (PI): Definition, Components, and Formula.* Available at: https://www.investopedia.com/terms/p/profitability.asp. [Accessed: 10 February 2023].

Fernando, J., 2022. *Net Present Value (NPV): What It Means and Steps to Calculate It.* Available at: https://www.investopedia.com/terms/n/npv.asp. [Accessed: 1 February 2023].

Fernando, J., 2022. *Working Capital: Formula, Components, and Limitations.* Available at: https://www.investopedia.com/terms/w/workingcapital.asp. [Accessed: 24 February 2023].

Hastak, M., 2015. *Skills & Knowledge of Cost Engineering,* 6th Edition. AACE International.

Hayes, A., 2022. *Return on Capital Employed (ROCE): Ratio, Interpretation, and Example.* Available at: https://www.investopedia.com/terms/r/roce.asp#:~:text=What%20Is%20Return%20on%20Capital,it%20is%20put%20to%20use. [Accessed: 20 February 2023].

Herold, T., 2014. *Financial Terms Dictionary.* Evolving Wealth, LLC.

Oxford, 2016. *Dictionary of Business and Management,* 6th Edition. Oxford University Press.

Tennent, J., 2013. *Guide to Financial Management,* 2nd Edition. The Economist in Association with Profile Books Ltd.

EXERCISE ANSWERS

1- 5 years.

2- Coal power plant 186% and solar power plant 136%.

   Second scenario → Coal power plant 157% and solar power plant 160%.

3- Complex ROI = [(20 + 30 + 50 + 60 + 80 + 80 + 60 + 50 + 50 + 40)/10]/(330 + 20) = 15%

4- ST Plus (32%) and YES Tecom (36%). ZKV Company (25%) and ALL Company (20%) do not achieve the target.

5- Break-even point when gas price achieves $ 6.4.

6- Option A NPV = $ 18,848,349. Option B NPV = $ 15,215,839. Option B has the lowest NPV.

7- The Table 6.12 shows the PI per project. The PI selection resulted in an over budget ($ 1,035 > budget), and some projects, such as Lion 03 and Dog 02, should be removed to respect the budget.

Project	Value	Selected
Lion 01	0.80	Yes
Lion 02	0.60	Yes
Lion 03	0.38	No
Mouse 01	0.55	Yes
Mouse 02	0.53	Yes
Dog 01	0.71	Yes
Dog 02	0.45	No
Cat 01	0.70	Yes
Cat 02	0.58	Yes

7 Economic Feasibility Study

The economic feasibility study is a crucial step in determining the project's financial viability, and it means analyzing how much profit the project is making. Also, the result should attend to the goals and benchmark.

The viability study can englobe many areas, such as environmental, technical, and quality – however, the chapter focuses only on the economic vision.

Figure 7.1 shows the process flow. Each step is explained in the following sections.

7.1 ASSUMPTIONS AND DATA ACQUISITION

The process starts with the data acquisition and assumptions definition. It means understanding clearly the scope or what should or should not be considered in the analysis. All assumptions should be registered and clear for the decision-makers. For example, if escalation is not considered in the analysis, the result can be more favorable than if this factor is included.

Also, the production and operation life should be realistic with the technology available because unrealistic assumptions can distort the result. One of the best practices is scenario analysis, where the key variables are changed to promote optimistic, realistic, and pessimist scenarios. It can generate various possible values, allowing decision-makers to choose the best choice aligned with the company strategy.

There is a list below with examples of assumptions and information that should be getting in this stage:

- Scope
- Asset life
- Operation period
- Production capacity
- Inflation
- Escalation
- Depreciation method
- Salvage value
- Interest rate per period
- Production price
- Cost estimation methodology

The scope should be cleared and registered. It ensures all stakeholders know the project objective, permitting future comparison and analysis. Asset life is the operation phase duration, generally in years. According to the technology available, it

DOI: 10.1201/9781003402725-7

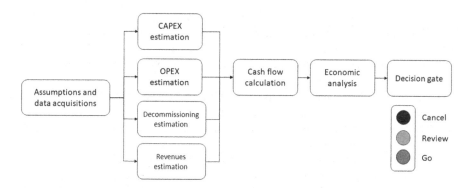

FIGURE 7.1   Process flow of the economic feasibility study

should be a realistic estimation because a very optimistic number can distort the study.

The operation period is the number of days or months that the asset will operate. Hence, shutdown days are not counted. The production capacity is a crucial assumption because it will affect all cost estimations and the plant size.

Inflation and escalation are discussed in Chapter 9, but they should be defined according to the country forecast where the asset will be built. The depreciation method and salvage value are discussed in the same chapter, and the technique should respect the current legislation.

Cost estimation methodology should be defined according to the planning phase. The conceptual methods are recommended for the initial stage, and the detailed method is when the project has a considerable maturity level.

Product price considers factors such as demand, offer, raw material cost, taxes, subsidies, and business strategy, but it is not the focus of this book.

All assumptions must be registered and defined carefully because they directly impact the study result.

Finally, special attention should ensure all factors use the correct unit and scale.

7.2   COST ESTIMATION

The second step is the cost estimation that should be done for the scope (assets) that compound the study.

The cost estimate step can be divided by the:

- OPEX estimation
- CAPEX estimation
- Decommissioning estimation
- Revenue estimation
- Depreciation estimation
- Tax estimation
- Gains estimation

The CAPEX, OPEX, and decommissioning estimations should be calculated by one of the methods discussed in Chapters 3, 4, and 5. For example, the deterministic value could be used for each cost estimate, or the accuracy range could be adopted for the scenarios, whereas the maximum value should be the pessimistic vision and the minimum, the optimistic scenario. Also, the probabilistic approach discussed in Chapter 5 could provide inputs for each scenario. For example, P25 could be the optimistic option, P50 the realistic, and P75 the pessimistic.

The revenue estimate is based on the product price and the production capacity. . Equation 7.1 estimates revenue:

Revenue = Price per product × plant capacity × operation period   (Equation 7.1)

The estimate should be done in a particular period, like year or day.
Depreciation methods are detailed in Chapter 9.

The tax estimation should be done according to the country's law. For this book's purpose, Equation 7.2 is used to estimate the tax:

Tax = Tax% × (revenue − opex − depreciation)   (Equation 7.2)

Finally, the gains or profit is estimated by Equation 7.3.

Gains = Revenue − OPEX − tax   (Equation 7.3)

7.3   CASH FLOW

The third step is the cash flow. It is the net amount of cash in and out of a company.

Figure 7.2 shows a cash flow schematic. The dashed arrows represent the inflows when cash is received, and the solid arrows the outflows when money is sent. Generally, the timeline is per year, and it is considered that the money spent and received is done in a unique moment per year.

7.4   ECONOMIC ANALYSIS

The next step is the economic analysis using the following indicators:

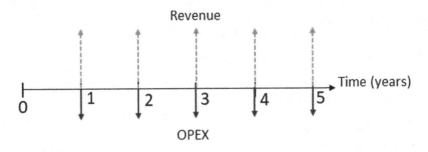

FIGURE 7.2   Cash flow diagram

- Payback period
- Return on investment (ROI)
- Return on capital employed (ROCE)
- Net present value (NPV)
- Profitability index (PI)
- Benchmarking
- Break-even analysis

After the indicator's calculation, a comparison is made between the results and goals or references. If the result is satisfactory, the project can advance to the next planning or execution phase.

7.5 DECISION GATE

The decision is made at a gate, and the board defines if the project should be reviewed, stopped, or go to the next planning phase. Usually, decision-makers analyze a report or dashboard, whereas scenarios, indicators, and a summary express the project's performance. The project can advance without comments, or they can request improvements, and the project is reviewed. Finally, the project could be canceled or postponed.

7.6 ECONOMIC FEASIBILITY STUDY – EXAMPLE 1

The example is based on an economic feasibility study for the initial project planning phase or conceptual phase. The project is a process plant of 20 tons/day capacity. The asset is estimated to operate for 20 years, 11 months per year. 1 month per year is the shutdown period for maintenance or operation issues.

The annual interest rate is 10%, and the tax is 15% on the gains. Also, this example is not considering escalation and inflation in the analysis.

These assumptions are listed in Table 7.1.

The assumptions allow us to move to the second stage and define the cost estimate. The study is the first planning phase. Hence, the quick and conceptual methods detailed in Chapter 3 should be used. The CAPEX cost estimate method is the capacity factor, and a similar plant has a cost of $ 159 million, a capacity of 43 tons/day, and an exponent 0.6 should be used. Hence, the result is $ 100.5 million.

$$\text{CAPEX (new plant)} = \$ 159 \text{ million} \times (20/43)^{0.6} = \$ 100.5 \text{ million}$$

The decommissioning cost estimate is defined using an industry reference which shows that the cost estimate is 40% of the CAPEX or $ 40.2 million..

In Chapter 3, the OPEX estimation is calculated by the parametric Equation 3.6, but in this example a different parametric model is proposed by Equations 7.4 and 7.5.

$$\text{OPEX} = \text{Raw material cost} + \text{maintenance cost} + \text{operation cost} + \text{others}$$

(Equation 7.4)

TABLE 7.1

Initial Assumptions for an Economic Feasibility Study – Example 1

Assumption	Value	Unit
Capacity	20	ton/day
Asset Life	20	years
CAPEX estimate method	Capacity factor	
CAPEX similar plant cost	159 million	$
Exponent (capacity factor)	0.6	
Operation time	11	Months per year
Operations days	335	per year
i = interest rate per period	10%	%
Tax %	15	% of gains
Depreciation method	Straight-line	-
Salvage value	5% of CAPEX	%
Inflation	Not included	%
Escalation	Not included	%
Product price	5,000	$ per ton

TABLE 7.2

OPEX Estimation – Example 1

Assumption	Value	Unit
Raw material cost per ton	1,000	US$/ton
Raw material cost per year	$1,000 \times 20 \times 335 = 6,700,000$	$/year
Maintenance cost	2,000,000	$/year
Operation cost	4,000,000	$/year
Other cost	1,000,000	$/year
OPEX	$3,350,000 + 2,000,000 + 4,000,000 + 6,700,000 =$ 13,700,000	$/year

Raw material cost = Cost per unit × plant capacity × operation period

(Equation 7.5)

The plant capacity and operation period are provided in Table 7.1. The assumption is to use historical information for maintenance and operation costs.

Examples of other costs are insurance, rent, research, and development.

The OPEX estimate is $ 13,700,000 per year, as listed in Table 7.2.

The following estimate is the revenue. A market study should be done to define the product price. The assumption for this study is $ 5,000 per ton. Using Equation 7.1, the revenue per year is:

Revenue = 5,000 × 20 × 335 = $ 33,500,000

The tax estimate is estimated using Equation 7.2, and the tax percentage is from Table 7.1. The depreciation methods are detailed in Chapter 9. The example adopts the straight-line method. Hence, the value is calculated by Equation 9.1.

Depreciation = (100,500,000 − 5,000,000)/20 years = 4,773,750 per year

And the tax is

Tax = 15% × (33,500,000 − 13,700,000 − 4,773,750) = 2,253,938 per year

The gains per year is estimated by Equation 7.3:

Gains = 33,500,000 − 13,700,000 − 2,253,938 = 17,546,063

Table 7.3 lists the cost estimates.

After that, a cash flow diagram should be elaborated. Figure 7.3 shows that the CAPEX is the first negative spend, then each year has a similar behavior: positive revenue and negative OPEX and tax. Finally, the last spend is decommissioning. The cash flow is calculated like below:

Cash flow = In − out
Cash flow = Revenue − (CAPEX + OPEX + tax + decommissioning)
Cash flow = 33,500,000 × 20 years − (100,500,000 + 13,700,000 × 20 years
+ 2,253,938 × 20 years + 40,200,000)
Cash flow = 670,000,000 − (459,778,750) = $ 210,221,250

Despite the positive cash flow results, the economic analysis should consider the time of value and the company's strategic goals to ensure the project's viability.

Also, the cash flow is helpful to show that NPV should get the present value of tax, OPEX, revenue, and decommissioning, which means that all "arrows" illustrated in Figure 7.3 should be "moved" to present (year 0) as discussed in Section 6.5, by NPV analysis to provide a better comparison.

TABLE 7.3
Cost Estimation – Example 1

Cost estimate	Value	Unit
CAPEX	100,500,000	$
Decomissioning	40,200,000	$
OPEX	13,700,000	$/year
Revenue	33,500,000	$/year
Depreciation	4,773,750	$/year
Tax	2,253,938	$/year
Gains	17,546,063	$/year

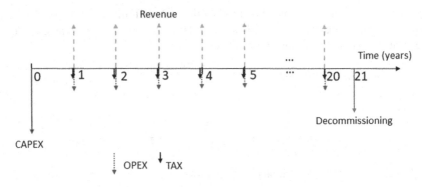

FIGURE 7.3 Cash flow diagram – Example 1 and icons made by Freepick from www.flaticon.com.

TABLE 7.4
Indicators and Goals – Example 1

Indicator	Description	Goal
PB	Payback period	< 6 years
ROI	Simple return on investment	> 200%
NPV	Net present value	> $ 40,000,000
PI	Profitability index	> 0.4

The economic analysis uses a group of indicators and company goals. They are listed in Table 7.4.

The goals ensure the project respects the company's objectives and can move to the next planning phase. For example, the business objective is to cover the investment under 6 years or PB < 6 years. Also, the NPV goal means that all cash, in and out during 20 years, at the present value, results in a profit above $ 40 million.

The indicators calculations are shown in Table 7.5, and their equations are discussed in Chapter 6. As illustrated in Figure 7.3, decommissioning is a single payment for the NPV calculation. And OPEX, tax, and revenue are a uniform series.

The last stage is the decision gate. Figure 7.4 summarizes the result showing that the project attends to all goals, allowing it to move to the next phase. However, the complete study should consider additional risks, environmental, regulatory, safety, and other factors.

7.7 ECONOMIC FEASIBILITY STUDY (ESCALATION INCLUDED) – EXAMPLE 2

Example 7.2 uses the assumptions and cost estimates of Example 1. The difference is that escalation is considered, and it aims to analyze the impact of these factors.

TABLE 7.5
Indicators Calculation – Example 1

	Indicator	Calculation	Result
	PB	PB = CAPEX/gains = 100,500,000/17,542,500	**5.7 years**
	ROI	ROI = (Gains × life – CAPEX)/CAPEX	**249%**
		ROI = (17,546,063 × 20 – 100,500,000)/100,500,000	
	Revenue	Uniform series = $((1 + i)^n - 1)/(i\,(1 + i)^n)$	285,204,384
		$33,500,000 \times (1 + 0.15)^{20} - 1/(0.15 \times (1 + 0.15)^{20})$	
	OPEX	Uniform series = $((1 + i)^n - 1)/(i\,(1+i)^n)$	116,635,823
		$13,700,000 \times (1 + 0.15)^{20} - 1/(0.15 \times (1 + 0.15)^{20})$	
	Tax	Uniform series = $((1 + i)^n - 1)/(i\,(1 + i)^n)$	19,189,041
NPV		$2,253,938 \times (1 + 0.15)^{20} - 1/(0.15 \times (1 + 0.15)^{20})$	
	Decom.	Single payment = $1/(1 + i)^n$	5,945,474
		$40,200,000 \times 1/(1 + 0.15)^{20}$	
	Result	NPV = Revenue – CAPEX – OPEX – Tax – Decom	**42,904,047**
		NPV = 285,204,384 – 100,500,000 – 116,635,823	
		– 19,189,041 – 5,945,474	
	PI	PI = NPV/CAPEX	**0.43**
		PI = 43,403,446/100,500,000	

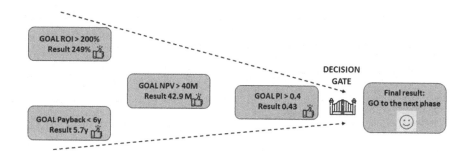

FIGURE 7.4   Decision gate – Example 1 and icons made by Freepick from www.flaticon.com.

Table 7.6 lists the escalation percentage per cost category during the operation life. The escalation percentage is a constant to simplify the calculation, but they are typically different per year and cost category.

Consequently, OPEX, raw material, revenue, tax, and gains will have different yearly values. The unique rubrics that are unaffected are CAPEX, depreciation, and commissioning cost.

Figures 7.5 and 7.6 show how OPEX, raw material, revenue, and gains change during the operation life.

The raw material is a component of OPEX, as in Equation 7.4, and it has a considerable increase of 4% per year, resulting in a 211% increment after 20 operation

TABLE 7.6
Escalation per Cost Category – Example 2

Cost categories	Escalation per year
Maintenance	3%
Operation cost	3%
Other cost	2%
Raw material	4%
Product Price	1%

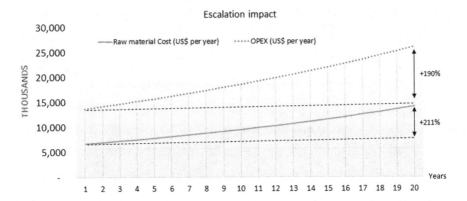

FIGURE 7.5   OPEX and raw material cost per year

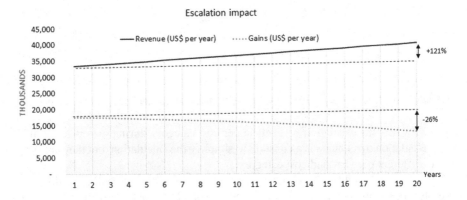

FIGURE 7.6   Revenue and gains per year

years. Also, the other categories that impact OPEX have an increment, resulting in 190% raised, as shown in Figure 7.5.

The product price escalation impacts the revenue 1% per year, which is lower than other cost categories' impacts. Hence, Figure 7.6 shows that revenue increased

121% in 20 years, which is insufficient to cover the increment from OPEX and tax spending. The result is that gains are reduced (– 26 %) along with the operation life, pushing the indicators to a worse outcome.

The raw material calculation for the first 3 years is shown below to exemplify the process per year.

The raw material cost per year, and the escalation % comes from Table 7.6.

Year 1 (from Example 1 – without escalation) = $ 6,700,000
Year 2 = 6,700,000 × (1 + 0.04) = $ 6,968,000
Year 3 = 6,968,000 × (1 + 0.04) = $ 7,246,720

Table 7.7 shows the result of the raw material, OPEX, revenue, and gain per year considering the escalation. Year 1 is the estimated value from Example 1.

Figure 7.7 shows the cash flow diagram where revenue and OPEX increase along with the operation life. However, tax decreases because the revenue increment (1% per year) is lower than the OPEX increase rate.

The cash flow is calculated like below:

Cash flow = In – out
Cash flow = Revenue – (CAPEX + OPEX + tax + decommissioning)
Cash flow = 737,636,634 – (100,500,000 + 385,032,743 + 38,569,334 + 40,200,000)
Cash flow = 737,636,634 – (564,302,076) = $ 173,334,557

Again the cash flow is positive but lower than the scenario without escalation (Example 1).

The economic analysis uses the same indicators and goals defined in Example 1.

The indicators calculations are shown in Table 7.8, and their equations are discussed in Chapter 6. The NPV results are shown in the Table 7.11.

The results are worse than the first example because of the escalation impact. Hence, the indicators do not achieve the goals, like in Figure 7.8.

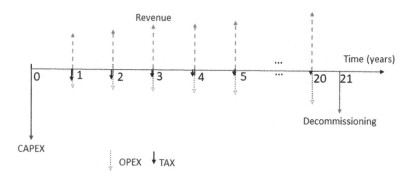

FIGURE 7.7   Cash flow – Example 2

TABLE 7.7
Raw Material Cost, OPEX, Revenue, and Gains per Year

Year	1	2	3	4	5	6	7
Raw material cost ($)	6,700,000	6,968,000	7,246,720	7,536,589	7,838,052	8,151,574	8,477,637
OPEX ($)	13,700,000	14,178,000	14,673,020	15,185,678	15,716,614	16,266,493	16,836,003
Revenue ($)	33,500,000	33,835,000	34,173,350	34,515,084	34,860,234	35,208,837	35,560,925
Gains ($)	17,546,063	17,433,013	17,308,768	17,172,848	17,024,755	16,863,969	16,689,952

Year	8	9	10	11	12	13	14
Raw material cost ($)	8,816,743	9,169,413	9,536,189	9,917,637	10,314,342	10,726,916	11,155,992
OPEX ($)	17,425,860	18,036,803	18,669,601	19,325,051	20,003,979	20,707,242	21,435,728
Revenue ($)	35,916,534	36,275,700	36,638,457	37,004,841	37,374,890	37,748,639	38,126,125
Gains ($)	16,502,146	16,299,969	16,082,818	15,850,068	15,601,067	15,335,141	15,051,587

Year	15	16	17	18	19	20	Total
Raw material cost ($)	11,602,232	12,066,321	12,548,974	13,050,933	13,572,971	14,115,889	199,513,126
OPEX ($)	22,190,360	22,972,093	23,781,919	24,620,867	25,490,002	26,390,432	387,605,748
Revenue ($)	38,507,386	38,892,460	39,281,385	39,674,198	40,070,940	40,471,650	737,636,634
Gains ($)	14,749,679	14,428,658	14,087,741	13,726,110	13,342,919	12,937,288	314,034,557

TABLE 7.8
Indicators Calculation – Example 2

Indicator		Calculation	Result
PB		PB = CAPEX/gains* = 100,500,000/17,224,902	**5.8 years**
ROI		ROI = (Gains** − CAPEX)/CAPEX	212%
		ROI = (314,034,557 − 100,500,000)/100,500,000	
NPV	Revenue	Single series = $1/(1 + i)^n$	304,710,985
		See Exercise Answers section	
	OPEX	Single series = $1/(1 + i)^n$	147,739,102
		See Exercise Answers section	
	Tax	Single series = $1/(1 + i)^n$	17,449,539
		See Exercise Answers section	
	Decom.	Single series = $1/(1 + i)^n$	5,975,473
		$40,200,000 \times 1/(1 + 0.15)^{20}$	
	Result	NPV = Revenue − CAPEX − OPEX − Tax − Decom.	**32,941,553**
		NPV = 304,710,985 − 100,000,000 − 148,450,532	
		− 17,373,154 − 5.945.745	
PI		PI = NPV/CAPEX	**0.33**
		PI = 32,941,553/100,500,000	

* Adopted the gains average of the six initial years, according to Table 7.7.
** Adopted the sum of gains to the operation life, according to Table 7.7.

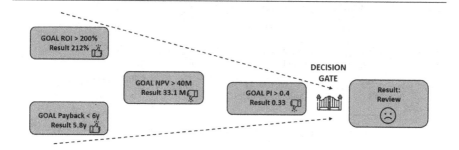

FIGURE 7.8 Decision gate – Example 2 and icons made by Freepick from www.flaticon.com

In addition, the assumption definition is crucial because an optimistic or pessimistic choice can be decisive. One solution is creating different scenarios.

7.8 ECONOMIC FEASIBILITY STUDY – EXERCISE 1

The manager asks to elaborate on an economic feasibility study for a project considering the assumptions and data below. Furthermore, a comparison and a recommendation should be made regarding the goals detailed below.

Assumptions and Data – Table 7.9
Goals – Table 7.10

TABLE 7.9
Initial Assumptions for an Economic Feasibility Study – Exercise 1

Assumption	Value	Unit
Capacity	20	ton/day
Asset life	30	years
Operation time	10.5	months
Operations days	320	per year
i = interest rate per period	12%	%
Tax %	16	% of gains
Depreciation method	Straight-line	-
Salvage value	5% of CAPEX	%
Escalation	Not included	%
Price per ton	10,000	US$/ton
CAPEX	200,000,000	US$
Decommissioning % of CAPEX	30%	%
Maintenance	3,000,000	US$/year
Operation cost	5,000,000	US$/year
Other cost	2,000,000	US$/year
Raw material	3,000	US$/ton

TABLE 7.10
Indicators and Goals – Exercise 1

Indicator	Description	Goal
PB	Payback period	< 6 years
ROI	Simple return on investment	> 200%
NPV	Net present value	> $ 40,000,000
PI	Profitability index	> 0.3

TABLE 7.11
Table below shows the NPV result per year of example 2

Year	1	2	3	4	5	6	7	8	9	10
Revenue	30,454,545	27,962,810	25,674,944	23,574,266	21,645,463	19,874,470	18,248,377	16,755,328	15,384,438	14,125,711
OPEX	12,454,545	11,709,091	11,008,655	10,350,494	9,732,034	9,150,856	8,604,694	8,091,417	7,609,029	7,155,653
Tax	2,049,034	1,846,271	1,661,955	1,494,485	1,342,396	1,204,344	1,079,099	965,538	862,631	769,436

Year	11	12	13	14	15	16	17	18	19	20	Total
Revenue	12,969,971	11,908,792	10,934,436	10,039,800	9,218,362	8,464,132	7,771,613	7,135,753	6,551,919	6,015,853	304,710,985
OPEX	6,729,528	6,329,002	5,952,523	5,598,634	5,265,967	4,953,238	4,659,242	4,382,846	4,122,987	3,878,668	147,739,102
Tax	685,091	608,809	539,869	477,613	421,440	370,798	325,186	284,146	247,258	214,140	17,449,539

BIBLIOGRAPHY

AACE International, 2022. *Recommended Practice 10S-90 Cost Engineering Terminology.* Available at: https://web.aacei.org/resources/cost-engineering-terminology. [Accessed: 8 January 2023].

Atrill, P., 2017. *Financial Management for Decision Makers*, 8th Edition. Pearson.

Chen, J., 2023. *Profitability Index (PI): Definition, Components, and Formula.* Available at: https://www.investopedia.com/terms/p/profitability.asp. [Accessed: 10 February 2023].

Defence Department, U.S., 2016. *Inflation and Escalation. Best Practices for Cost Analysis.* Available at: https://www.cape.osd.mil/files/InflationandEscalationBestPracticesforCostAnalysisforWebsiteForPubRelReview.pdf. [Accessed: 1 April 2023].

Drury, C., 2015. *Management and Cost Accounting – Student Manual*, 9th Edition. Cengage Learning.

Fernando, J., 2022. *Net Present Value (NPV): What It Means and Steps to Calculate It.* Available at: https://www.investopedia.com/terms/n/npv.asp. [Accessed: 1 February 2023].

Hastak, M., 2015. *Skills & Knowledge of Cost Engineering*, 6th Edition. AACE International.

Herold, T., 2014. *Financial Terms Dictionary.* Evolving Wealth, LLC.

Oxford, 2016. *Dictionary of Business and Management*, 6th Edition. Oxford University Press.

Tennent, J., 2013. *Guide to Financial Management*, 2nd Edition. The Economist in Association with Profile Books Ltd.

Whiteley, J., 2004. *Financial Management.* Palgrave Master Series.

EXERCISE ANSWER

1- Payback period = 6.61 years
   Simple ROI = 354%
   NPV = 41.6 million
   Profitability index (PI) = 0.21

The project does not achieve the payback period and PI goals. One recommendation could be a CAPEX cost optimization.

8 Cost Control

The cost control aims to monitor the project performance and generate analysis and diagnosis. AACE International, 2022, defines cost control as *"the application of procedures to monitor expenditures and performance against the progress of projects or manufacturing operations; to measure variance from authorized budgets and allow effective action to be taken to achieve minimum costs."*[1]

There are several techniques to control the cost, and the chapter discusses the following methods:

- Planned × performed
- Earned value management
- Rundown curve
- S curve
- Indicators

Cost control is done periodically (e.g., monthly) and can show different project views. For example, project performance could be measured from a perspective of the lifecycle, current year, quarter, or month. The visions adopted are according to the business necessity and cost management plan.

The cost management plan shows the cost control periodicity, responsibilities and rules, level of details for each cost category, and methodology to execute the control.

The challenges are to ensure that data is organized and reliable. Data should be normalized, if necessary, but never frauded or disguised to hide the actual scenario, because transparency is essential for actions to be taken at the right time. In addition, tools and software should be used to reduce manual activities, improving automation and analysis time.

Finally, the cost control details should be according to the audience. For example, a high-level report does not analyze the WBS's last level. On the contrary, they should be able to evaluate the deliverable's performance. Also, the cost control report should be documented and available for future audits.

8.1 PLANNED × PERFORMED

The method is based on the planned, performed cost and forecast comparisons. Overall, the planned cost is approved for a work package or activity. The sum of all work packs or activities results in the total cost estimate for a project. The planned cost for house construction in Figure 8.1 is $ 100,000 and $ 500,000 for the project.

Also, the planned cost is expected to be the project baseline, as discussed in Chapter 4. Although the baseline is expected not to change, modifications are a reality in many projects in real life. Hence, when a variation occurs – for example, an additional scope – a change process should be applied to ensure that the claim is

DOI: 10.1201/9781003402725-8

FIGURE 8.1 Planned cost per house and total planned cost example and icons made by Freepick from www.flaticon.com

FIGURE 8.2 Baseline changes process and icons made by Freepick & Parzival from www.flaticon.com

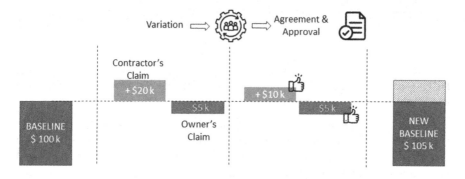

FIGURE 8.3 Baseline variation example and icons made by Freepick from www.flaticon.com

analyzed, negotiated, approved, documented, and communicated to all partners. Figure 8.2 summarizes this process.

Figure 8.3 shows an example. The baseline is $ 100,000, but there are claims from the contractor ($ + 20,000) and the owner ($ – 5,000). After the change management process, the contractor claim is reduced to + 10,000, and the owner claim is accepted (scope reduction). Hence, the new baseline is $ 105,000.

The second comparison parameter is the performed cost or actuals, which means the actual expenditures incurred by a program or project. For example, the houses illustrated in Figure 8.4 have a planned cost of $ 500,000; after the accomplishment, the total performed/actual cost is $ 600,000. In this case, there is an overrun when the actual cost exceeds the planned cost. If the opposite occurs, there is an underrun.

The forecast is the last parameter analyzed. According to AACE International, 2022, it *"estimates and predicts future conditions and events based on information and knowledge available during the forecast."*[2]

FIGURE 8.4 Actuals (or performed) and planned costs example and icons made by Freepick from www.flaticon.com

FIGURE 8.5 Planned, performed, and forecast for the houses construction project and icons made by Freepick from www.flaticon.com

For example, after three houses are completed, the forecast for the last two houses is $ 110,000 per house. The contractor expects that the learning curve will decrease compared to performed cost. Hence, the project completion estimate is the actual cost plus the forecast, $580,000, such as in Figure 8.5.

Several methods estimate the forecast, such as expert opinion, mathematical model, planned, or a new estimate. The expert opinion is based on the estimation done by a specialist with high experience in similar activities or projects. The mathematical method is discussed in Example 8.1. The planned method assumes that the forecast will be equal to the planned cost, and, finally, a new calculation could be done to define the forecast.

8.1.1 Planned × Performed – Graph Analysis – Example 1

The example aims to show the graph analysis to support cost control.

Project ABC has a lifecycle of 2 years, and it has two phases: planning (1 year) and execution phase (1 year). The planned cost is $ 1,300 for the planning phase, $ 14,000 for the execution phase, and the total planned cost is $ 15,300, as shown in Figure 8.6.

Also, the left graph shows the cumulative curve, meaning that the planned monthly value is added cumulatively. And the right stacked column graph shows the total per year.

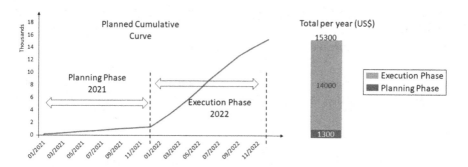

FIGURE 8.6   Planned Cost – Lifecycle – Example 1

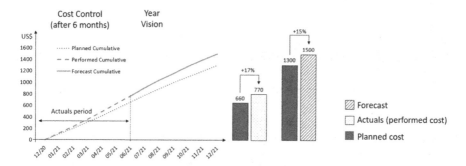

FIGURE 8.7   Planned × Performed Cost – Current Year – Example 1

To illustrate the method, a comparison between the planned and performed is made after 6 months, as shown in Figure 8.7 and Table 8.1, during the planning phase. Also, the analysis focuses on the current year view, 2021, and not the lifecycle perspective. The forecast is estimated by the project management team.

The performed cost up-to-date is $ 770 versus a planned cost of $ 660, resulting in a variation of $ 110 or 17%. In addition, a forecast cumulative curve is plotted, showing that the expectation was an overrun cost of 15% in 2021. Hence, the question is, why?

Figures 8.8 and 8.9 provide graphs to explain the variance in 2021. First, the waterfall graph provides a breakdown of the 2021 planned cost to the forecast. Scope changes, redesign/rework, and exchange rate promote the higher cost, and unfortunately, higher productivity is not enough to mitigate these costs. This information could be useful in creating a plan to minimize the rework and reduce the scope changes.

The second radar graph shows higher variances because the further away from the centre (0%), the more significant contribution to the deviation. It defines which area (department) is responsible for the variation. This information helps determine what actions can be taken to mitigate the cost overrun. The discipline's variance is shown in Table 8.2, and Figure 8.9 shows that civil construction and mechanical are responsible for 85% of the variance.

TABLE 8.1
Planned Cost, Performed (Actual), and Forecast

Month	Planned	Planned cumulative	Performed	Performed cumulative	Forecast
30/01/2021	100	100	115	115	-
28/02/2021	100	200	120	235	-
30/03/2021	110	310	125	360	-
30/04/2021	110	420	130	490	-
30/05/2021	120	540	140	630	-
30/06/2021	120	660	140	770	-
30/07/2021	120	780			140
30/08/2021	110	890			130
30/09/2021	110	1,000			120
30/10/2021	100	1,100			120
30/11/2021	100	1,200			110
30/12/2021	100	1,300			110

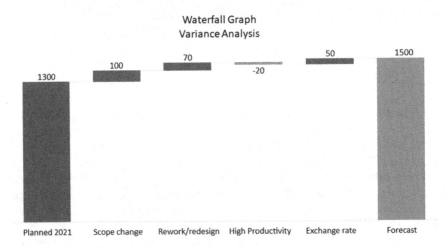

FIGURE 8.8   Variance analysis by waterfall graph – Example 1

8.1.2   Forecast – Mathematical Model – Example 2

The mathematical model starts with data analysis to determine the best math method, which could be simple statistic calculations, such as average and median, or mathematical models, such as trend curves. Finally, the forecast should be estimated, and a coherent test should be done. The sequence is summarized in Figure 8.10.

The project CBA has 6 months of duration, and planned and performed costs are listed in Table 8.3. The total planned cost is $ 1,200. What is the forecast for the sixth month?

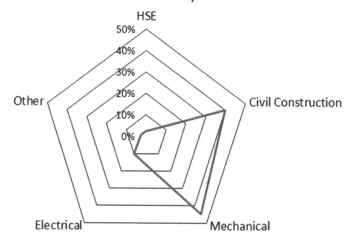

FIGURE 8.9   Variance analysis by radar graph – Example 1

TABLE 8.2
Variance by Discipline

Healthy, safety, and environemt (HSE)	Civil construction	Mechanical	Electrical	Other
3%	40%	45%	10%	3%

Data analysis ➡ Math method choice ➡ Application and test

FIGURE 8.10   Mathematical model sequence

TABLE 8.3
Planned and Performed Cost – Example 2

Month	1	2	3	4	5	6
Planned ($)	200	200	200	200	200	200
Performed ($)	100	180	120	150	140	

The assumption is that following the expertise and parameters of similar projects, the average applies to it. Hence the performed average is calculated and adopted as the forecast:

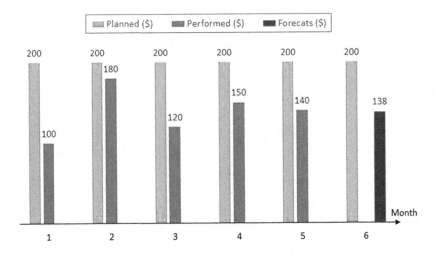

FIGURE 8.11   Planned, actuals (or performed) and forecast – Example 2

$$\text{Forecast} - \text{Month } 6 \frac{\left(100 + 180 + 120 + 150 + 140\right)}{5} = \$ 138$$

Hence the project's forecast is $100 + 180 + 120 + 150 + 140 + 138 = \$ 828$

The forecast is shown in Figure 8.11, and the result is more realistic than the planned value. This type of method has some benefits because it is straightforward to apply and helps to challenge other methods as well. However, it should be used carefully because the calculation can generate distortions if applied, for example, in an earlier project's stage.

8.1.3   PLANNED × PERFORMED – EXAMPLE 3

The Sun Company has five projects in the current year and a budget of $ 29.1 million to execute them. The budget for each project is listed below:

Blue Sky Project –	Planned $ 1.87 million
Black Sea Project –	Planned $ 3.98 million
Green Garden Project –	Planned $ 5.60 million
White Building Project –	Planned $ 6.05 million
Yellow Road Project –	<u>Planned $ 11.6 million</u>
Portfolio –	**Planned $ 29.1 million**

After 8 months, it is questioned whether the budget is enough. The cost driver is not expenditure more than CAPEX planned. Also, the analysis is in the current year (not for the project lifecycle).

Table 8.4 shows the planned and actuals per project and the portfolio, which is the sum of the projects. Also, the table shows the forecast provided by each project

TABLE 8.4
Planned, Performed (Actuals until August), Forecast, and Forecast Math Model per Project and Portfolio (Total) – Example 3

$ Thousand		1	2	3	4	5	6	7	8	9	10	11	12
Blue Sky Project	Planned	100	110	132	140	144	162	163	175	176	185	188	195
	Performed	80	100	110	120	122	130	134	136				
	Forecast									140	145	150	155
	Forecast – Math									133	133	133	133
Black Sea Project	Planned	250	260	250	300	320	400	500	550	400	300	250	200
	Performed	240	240	250	250	280	320	340	380				
	Forecast									700	350	250	200
	Forecast – Math									347	347	347	347
Green Garden Project	Planned	320	352	444	500	580	550	500	480	475	465	470	464
	Performed	250	300	310	335	350	390	400	400				
	Forecast									405	410	412	415
	Forecast – Math									397	397	397	397
White Building Project	Planned	321	310	340	382	435	445	512	575	605	648	710	767
	Performed	325	884	393	496	593	623	595	645				
	Forecast									462	450	440	413
	Forecast – Math									621	621	621	621
Yellow Road Project	Planned	520	560	620	710	780	1,100	1,100	1,157	1,150	1,159	1,289	1,455
	Performed	520	615	710	795	910	1,210	1,290	1,350				
	Forecast									1,150	1,159	1,289	1,455
	Forecast – Math									1,283	1,283	1,283	1,283
Portfolio	Planned	1,511	1,592	1,786	2,032	2,259	2,657	2,775	2,937	2,806	2,757	2,907	3,081
	Performed	1,415	2,139	1,773	1,996	2,255	2,673	2,759	2,911				
	Forecast									2,857	2,514	2,541	2,638
	Forecast – Math									2,781	2,781	2,781	2,781

manager and a second forecast based on the average of the last three performed months.

In addition, the report should answer the following questions:

What is each project variation in the current month? And up-to-date? And in the forecast?

What is the portfolio variation in the current month? And in the forecast? And up-to-date?

Is the forecast reliable through the mathematical forecast (from 3 months' average)?

Which project should be challenged through the mathematical model (last 3 months)?

The answer is divided into three periods: current month (August), up to date, and the year.

Current Month: August. The analysis is shown in Table 8.5.

The portfolio has a minor variance, 0.9%, but there are attention points. Three projects have positive variances showing that they spent under the plan. On the other hand, two projects, White Building and Yellow Road, are cost overrun and compensate for the others' underspend. The Yellow Road, which has the highest planned in the period, $ 1,157, spent 14.3% over the planned, which is the highest variance.

Consequently, despite the good portfolio performance, the projects must be justified, and variances explained to mitigate these variances.

Up-to-Date. The analysis is shown in Table 8.6.

The up-to-date means the cumulative result from January to August. The portfolio has been spent over, and it becomes a yellow signal. The portfolio must respect the

TABLE 8.5
Planned, Performed Value on August, and Variance in % and $

		Month 8	Variance $ Planned – performed	Variance % (Planned / Perf.) – 1
	$ Thousand			
Blue Sky Project	Planned	175	39	28.7
	Performed	136		
Black Sea Project	Planned	550	170	44.7
	Performed	380		
Green Garden Project	Planned	480	80	20
	Performed	400		
White Building Project	Planned	575	– 70	10.9
	Performed	645		
Yellow Road Project	Planned	1,157	– 193	14.3
	Performed	1,350		
Portfolio	Planned	2,937	26	0.9
	Performed	2,911		

TABLE 8.6
Planned, Performed Value Up-to-Date, and Variance in % and $

$ Thousand		Total up-to-date	Variance $ Planned – Performed	Variance % (Planned / Perf.) – 1
Blue Sky Project	Planned	1,126	194	20.8
	Performed	932		
Black Sea Project	Planned	2,830	530	23.0
	Performed	2,300		
Green Garden Project	Planned	3,726	991	36.2
	Performed	2,735		
White Building Project	Planned	3,320	−1,234	−27.1
	Performed	4,554		
Yellow Road Project	Planned	6,547	−853	−11.5
	Performed	7,400		
Portfolio	Planned	17,549	− 372	−2.1
	Performed	17,921		

year budget, which is not current. The two major budget projects, White Building and Yellow Road, have been cost overrun, like in the current period. They are − 27.1% and − 11.5% over the plan, respectively, and the portfolio result is not worse because the others are underspent.

Year – Forecast. The analysis is shown in Table 8.7.
The Blue Sky Project forecast is shown below to exemplify the forecast math:

$$\text{Forecast Math method: } (136 + 134 + 130) / 3 = 133$$

The forecast based on the math model shows that White Building Project should be challenged. It is the second project that most impacted the portfolio results, and the math model shows a − 14.0% variance versus − 4.3%, according to the project manager. Also, Yellow Road Project even shows an overspend forecast for both methods confirming the trends up-to-date. On the other hand, for the three projects which the performance is under planned, the forecast shows the same result, but Black Sea and Green Garden have an improvement trend, and Blue Sky is increasing the variance compared to the up-to-date data.

Portfolio forecast provided by the project management is under the budget, but just 2.2% means any variation can impact it. What supports this concern is that u-to-date vision, the portfolio has not respected the planned value. Also, the forecast based on a math model which aims to challenge the project management forecast shows that the estimated budget will be almost exceeded, just 0.2% below. It means that all attention should be paid to the last quarter.

Finally, the projects must be justified, variances explained, and action plans created. Furthermore, some activities could be postponed to avoid a portfolio being over budget.

TABLE 8.7

Planned, Performed, Forecast, and Forecast Math Model in Year Vision, and Variance in % and $

$ Thousand		Total Planned	Variance $ Plan. – Forecast	Variance % (Plan./ Forec.) – 1	Variance $ Plan – Forec. Math	Variance % (Plan. – F. Math) – 1
Blue Sky Project	Planned	1,870	348	22.9	405	27.6
	Performed	932				
	Forecast	1,522				
	Forecast Math Model	1,465				
Black Sea Project	Planned	3,980	180	4.7	– 293	8.0
	Performed	2,300				
	Forecast	3,800				
	Forecast Math Model	3,687				
Green Garden Project	Planned	5,600	1223	27.9	1278	29.6
	Performed	2,735				
	Forecast	4,377				
	Forecast Math Model	4,322				
White Building Project	Planned	6,050	– 269	– 4.3	– 988	–14.0
	Performed	4,554				
	Forecast	6,319				
	Forecast Math Model	7,038				
Yellow Road Project	Planned	11,600	– 853	– 6.8	– 933	–7.4
	Performed	7,400				
	Forecast	12,453				
	Forecast Math Model	12,533				
PORTFOLIO	Planned	29,100	629	2.2	55	0.2
	Performed	17,921				
	Forecast	28,471				
	Forecast Math Model	29,045				

TABLE 8.8
EVM System Terms

System terms	Description	Meaning
BCWS or PV	Budgeted cost for work scheduled or planned value	What is planned to be done
BCWP or EV	Budgeted cost for work performed or earned value	What was done
ACWP or AC	Actual cost of work performed or actual cost	The cost incurred

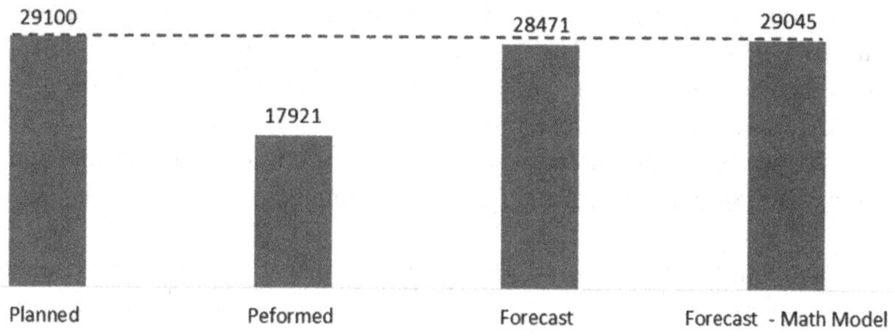

FIGURE 8.12   Planned, performed, forecast, and forecast by math model of the portfolio in the year vision

Figure 8.12 summarizes the portfolio planned (budget), performed (actuals), and forecasts. The dashed line shows the portfolio budget limit.

8.2   EARNED VALUE MANAGEMENT

Earned value management is one of the most helpful and famous project control methods. It is defined as *"a project progress control system that integrates work scope, schedule, and resources to enable objective comparison of the earned value to the actual cost and the planned schedule of the project"*[3] (AACE International, 2022).

The method is based on three terms. The budgeted cost for word scheduled (BCWS), or planned value (PV), is what is planned to be done. Then, the budgeted cost for work performed (BCWP), or earned value (EV), is what is done. Finally, the actual cost of work performed (ACWP), or actual cost (AC), is the cost incurred, as summarized in Table 8.8.

One difference from the previous method is the earned value (EV) concept or budgeted cost for work performed (BCWP). It is defined as a *"measure of the value of work performed so far. The 'value' of the work earned at the date of analysis (data date)"* (AACE International, 2022).[4] It is important to highlight that EV and AC are not the same. EV means how much the work that was actually performed should have been done. In comparison, AC is the actual (real) cost of the work done.

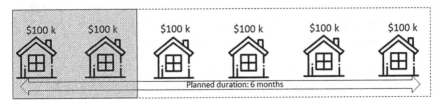

PV Per house: $ 100 k
Project Planned Value: $ 600 k
▨ Scope done after 3 months

FIGURE 8.13   Earned value example and icons made by Freepick from www.flaticon.com

The following example is to illustrate the EV concept. Considering a scope to build six houses, which the PV is $ 600,000 and duration is 6 months. After 3 months, two homes are made, and the initial question is, what is the EV?

Figure 8.13 summarizes the example.

The answer is $ 200,000 because two houses are built in the period, and PV per house is $ 100,000. Overall the EV value is estimated by Equation 8.1:

$$\text{Earned value (EV)} = \% \text{ complete} \times \text{budget for that account} \qquad \text{(Equation 8.1)}$$

Then, the calculation is:

Two of six houses times the project's budget, or 2/6 × 600,000, and the result is $ 200,000. However, what does it mean? Is the project cost overrun or not? And is the schedule behind or not? These questions are answered by the cost variance (CV), schedule variance (SV), cost performance index (CPI), and schedule performance index (SPI).

Cost Variance (CV)

The cost variance is calculated by the difference between the earned value and actual cost, as shown in Equation 8.2.

$$\text{Cost variance (CV)} = \text{EV} - \text{AC} \qquad \text{(Equation 8.2)}$$

If the CV is negative, the actual cost is higher than the earned value. Hence, the cost performance is poor, and there is an overrun. On the other hand, a positive CV shows that AC is lower than EV, which means that the performance is better than planned. Figure 8.14 summarizes the possible CV results.

For example, if the actual cost (AC) is $ 300,00 after two houses are built, the CV estimated is:

$$CV = 200,000 - 300,000 = \$ -100,000$$

It shows that the project is cost overrun, and a mitigation plan must be implemented.

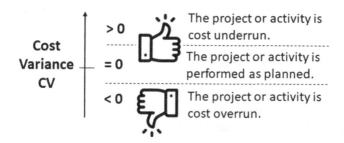

FIGURE 8.14 Cost variance meaning and icons made by Freepick from www.flaticon.com

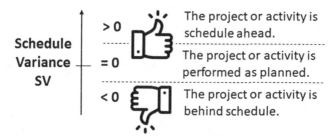

FIGURE 8.15 Schedule variance meaning and icons made by Freepick from www.flaticon.com

Schedule Variance (SV)

The scheduled variance is calculated by the difference between the earned value and planned value, as shown in Equation 8.3.

$$\text{Schedule variance (SV)} = EV - PV \qquad \text{(Equation 8.3)}$$

A positive result shows the project or activity is scheduled ahead. If the SV is negative, the project or activity is behind schedule, and zero means it is according to plan. Figure 8.15 summarizes the possible SV results.

Through the same example of building six houses, the planned value (PV) after 3 months is $ 300,000, so the SV is:

$$SV = 200,000 - 300,000 = \$ -100,000$$

It shows that the project is behind schedule, and actions must be taken to promote a correction.

Cost Performance Index (CPI)

The cost performance index (CPI) is calculated by the ratio between the earned value and actual cost, as with Equation 8.4. The CV provides an absolute variance measure, whereas the CPI shows a relative variance.

$$\text{Cost Performance Index } \left(CPI\right) = \frac{EV}{AC} \qquad \text{(Equation 8.4)}$$

The possible results for a CPI are shown in Figure 8.16. Although a CPI higher than one should be read as a good result, if the number is so high, for example, four, it can indicate that the planning was too conservative.

The CPI for the houses construction project is:

$$CPI = 200{,}000/300{,}000 = 0.67$$

The result is aligned with the CV, which shows that the project is spending more than planned. The overrun reasons should be identified and mitigated.

Schedule Performance Index (SPI)

The schedule performance index (SPI) is calculated by the ratio between the earned value and planned, as Equation 8.5. The SV provides an absolute variance measure. Meanwhile, the SPI shows a relative variance.

$$\text{Schedule Performance Index } (CPI) = \frac{EV}{PV} \qquad \text{(Equation 8.5)}$$

The possible results for an SPI are shown in Figure 8.17. Similar to the CPI, the performance is as planned if the result equals one. Over one, the project or activity is behind schedule, or there is a float. Also, if the result is under one, action should be taken because the project or activity is behind schedule.

The SPI for the houses construction project is calculated below:

$$SPI = 200{,}000/300{,}000 = 0.67$$

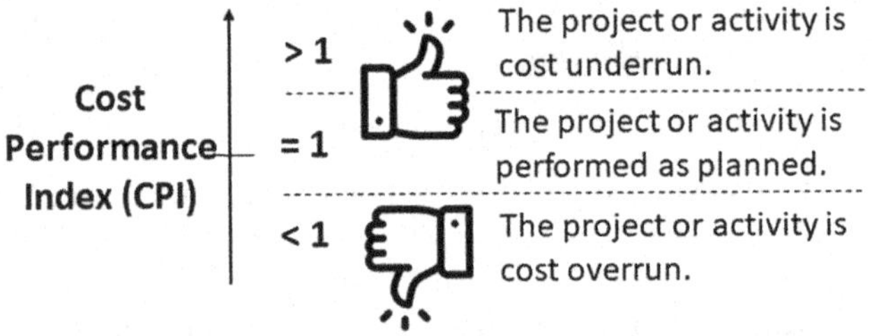

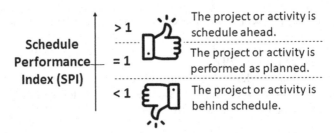

FIGURE 8.16 Cost performance index meaning and icons made by Freepick from www.flaticon.com

FIGURE 8.17 Schedule performance index meaning and icons made by Freepick from www.flaticon.com

Hence, the project is behind schedule, SPI = 0.67, and cost overrun, CPI = 0.67. Both, cost and schedule must be analyzed, revised, and correction actions implemented.

8.2.1   EVM APPLICATION – EXAMPLE 4

The process pilot plant aims to test a solution before implementing a large-scale new technology. The project has a budget of $ 31,025,000, and the duration is one year, starting in January. Tables 8.9 (current period) and 8.10 (up-to-date) list the planned and actual cost and % of completion per discipline.

After 9 months (September) an EVM report is requested, showing the earned value, schedule and cost variances, schedule, and cost performance index. Also, two visions should be provided: the current period and cumulative.

The variations are estimated by Equations 8.2 and 8.3 and indexes by 8.4 and 8.5. The data are listed in Table 8.9 for the current period and Table 8.10 for the cumulative period. The civil construction and piping estimation are shown below to exemplify the calculation of the current period:

Construction Civil

1) Earned Value (EV)
   EV = Budget × % completion = 4,100,000 × 5% = $ 205,000.
2) Cost variance (CV)
   CV = EV – AC = 205,000 – 195,100 = $ 9,900
3) Schedule variance (SV)
   SV = EV – PV = 205,000 – 50,000 = $ 155,000
4) Cost performance index (CPI)
   CPI = EV/AC = 205,000/195,100 = 1.05
5) Schedule performance index (SPI)
   SPI = EV/PV = 205,000/50,000 = 4.10

TABLE 8.9
Current Period Data

| | Current period (September) | | | |
Discipline	Total planned $ (year)	% Complete	Planned value (PV) $	Actual cost (AC) $
Project management	1,225,000 .	8	102,083	104,500
Design	1,800,000	0	-	-
Civil construction	4,100,000	5	50,000	195,100
Piping	7,000,000	25	2,100,000	2,098,000
Mechanics	9,800,000	30	3,430,000	3,100,000
Electrical and instrumentation	4,600,000	15	460,000	710,050
Commissioning	2,500,000	0	-	-
Total	31,025,000	18	6,142,083	6,207,650

TABLE 8.10
Up-to-Date Data

	Cumulative: Up-To-Date			
Discipline	Total planned $ (year)	% Complete	Planned value (PV) $	Actual cost (AC) $
Project management	1,225,000	75	918,750	930,500
Design	1,800,000	100	1,800,000	1,900,000
Civil construction	4,100,000	95	4,100,000	4,090,100
Piping	7,000,000	40	3,500,000	3,350,000
Mechanics	9,800,000	45	5,390,000	4,500,300
Electrical and instrumentation	4,600,000	15	690,000	650,000
Commissioning	2,500,000	0	-	-
Total	31,025,000	47	16,398,750	15,420,900

Initially, the construction civil result is considered good because SV and CV are positive. Hence, CPI and SPI are above one, as well. However, the SPI, 4.10, is so high and has become an attention point. It must be investigated, and cumulative vision could provide more details.

Piping

1) Earned value (EV)
   EV = Budget × % completion = 7,000,000 × 25% = $ 1,750,000
2) Cost variance (CV)
   CV = EV − AC = 1,750,000 − 2,098,000 = $ − 348,000
3) Schedule variance (SV)
   SV = EV − PV = 1,750,000 − 2,100,000 = $ − 350,000
4) Cost performance index (CPI)
   CPI = EV/AC = 1,750,000/2,098,000 = 0.83
5) Schedule performance index (SPI)
   SPI = EV/PV = 1,750,000/2,100,000 = 0.83

The piping results show that CV and SV are negative, which means a performance worse than planned and a cost overrun. The indicators CPI, 0.83, and SPI, 0.83, under one, confirm that a mitigation plan should be implemented for this discipline.

Table 8.11 shows the current period's results for each discipline and project. Overall, the project is behind schedule, SPI = 0.93 and cost overrun, CPI = 0.92.

Attention points are piping and mechanics because of the higher deviations, CV and SV, as well the lower indicator's results, CPI and SPI.

Project management has the best performance because it is a stable activity. Design is not analyzed because there is no planned and performed in the current period.

TABLE 8.11
Current Period Results

Discipline	Total planned $ (year)	% Complete	EV $	PV $	AC $	CV $	SV $	CPI	SPI
PM	1,225,000	8	102,083	102,083	104,500	(2,417)	-	0.98	1.00
Design	1,800,000	0	-	-	-	-	-	N/A	N/A
Civil Const.	4,100,000	5	205,000	50,000	195,100	9,900	155,000	1.05	4.10
Piping	7,000,000	25	1,750,000	2,100,000	2,098,000	(348,000)	(350,000)	0.83	0.83
Mec.	9,800,000	30	2,940,000	3,430,000	3,100,000	(160,000)	(490,000)	0.95	0.86
Elect. & Inst.	4,600,000	15	690,000	460,000	710,050	(20,050)	230,000	0.97	1.50
Commi.	2,500,000	0	-	-	-	-	-	N/A	N/A
Total	31,025,000	18	5,687,083	6,142,083	6,207,650	(520,567)	(455,000)	0.92	0.93

The same disciplines are analyzed in the cumulative period, from January to September, as below.

Construction Civil

6) Earned value (EV)
   EV = Budget × % completion = 4,100,000 × 95% = $ 3,895,000.
7) Cost variance (CV)
   CV = EV − AC = 3,895,000 − 4,090,100 = $ − 195,100
8) Schedule variance (SV)
   SV = EV − PV = 3,895,000 − 4,100,000 = $ − 205,000
9) Cost performance index (CPI)
   CPI = EV/AC = 3,895,000/4,090,100 = 0.95
10) Schedule performance index (SPI)
   SPI = EV/PV = 3,895,000/4,100,000 = 0.95

Cumulative vision shows that construction civil should be finished in September because the PV value equals the project budget, $ 4,100,000. However, the % completion and earned value show that 5% still needs to be concluded. For this reason, the SV is negative, $ − 205,000, and SPI is under one, 0.95. Also, the cost performance is the opposite of the current period. Cost variance is negative, $ − 195,000, and CPI under one, 0.95.

As a result, there is a delay, and the discipline has a cost overrun.

Piping

11) Earned value (EV)
   EV = Budget × % completion = 7,000,000 × 40% = $ 2,800,000
12) Cost variance (CV)
   CV = EV − AC = 2,800,000 − 3,350,000 = $ − 550,000
13) Schedule variance (SV)
   SV = EV − PV = 2,800,000 − 3,500,000 = $ − 700,000
14) Cost performance index (CPI)
   CPI = EV/AC = 2,800,000/3,350,000 = 0.84
15) Schedule performance index (SPI)
   SPI = EV/PV = 2,800,000/3,500,000 = 0.80

The piping result shows that discipline has a similar performance in cumulative vision to the current period. Hence, it is behind the schedule, SV = $ 700,000 and SPI = 0.80, and has a cost overrun, CV = $ − 550,000 and CPI = 0.84. The bias can confirm that actions should be taken because the discipline is struggling to achieve the plan.

Table 8.12 shows the cumulative vision results for each discipline and project from January to September.

TABLE 8.12
Cumulative Results

Discipline	Total planned $ (year)	% Complete	EV $	PV $	AC $	CV $	SV $	CPI	SPI
PM	1,225,000	75	918,750	918,750	930,500	(11,750)	-	0.99	1.00
Design	1,800,000	100	1,800,000	1,800,000	1,900,000	(100,000)	-	0.95	1.00
Civil const.	4,100,000	95	3,895,000	4,100,000	4,090,100	(195,100)	(205,000)	0.95	0.95
Piping	7,000,000	40	2,800,000	3,500,000	3,350,000	(550,000)	(700,000)	0.84	0.80
Mec.	9,800,000	45	4,410,000	5,390,000	4,500,300	(90,300)	(980,000)	0.98	0.82
Elect. and Inst.	4,600,000	15	690,000	690,000	650,000	40,000	-	1.06	1.00
Commi.	2,500,000	0	-	-	-	-	-	N/A	N/A
Total	31,025,000	47	14,513,750	16,398,750	15,420,900	(907,150)	(1,885,000)	0.94	0.89

Positive performance:

Project management has an actual like planned, CPI = 0.99 and SPI = 1.

Design performance is excellent from a time perspective, SPI = 1, and a reasonable cost result, CPI = 0.95.

Electrical and instrumentation is in the initial stage, 15% of completion, but shows an excellent result, CPI = 1.06 and SPI = 1.

Negative performance:

Construction civil and tubulation were discussed previously.

Mechanics has a good cost performance, CPI is close to 1 (CPI = 0.98), but the schedule is the attention point, SPI = 0.82.

Overall, the project deserves attention because it has a cost variation of $ − 907,150, CPI = 0.94, and schedule variation of $ − 1,885,000, SPI = 0.89. Also, the planned duration is 12 months, meaning that 47% of completion was executed in 9 months, and there are 3 months to finish 53%.

8.2.2 Graph Analysis

The graph analysis promotes a quick project examination. A typical visualization is a cumulative curve, as shown in Figure 8.18. The solid curve shows the planned value (PV), the dotted curve represents the actual cost (AC), and the dashed curve is the earned value (EV).

Through assessment date, it is possible to visualize the cost and schedule deviations. The difference between the dashed and solid curves is the schedule variance, SV, and between the dashed and dotted curves is the cost variance, CV.

Also, it is simple to identify if there is a cost overrun or underrun. If the earned value curve is under the actual curve, like EV curve 1 in Figure 8.19, the activity or project has a cost overrun. On the other hand, if the earned value curve is above the actual curve, there is a cost underrun.

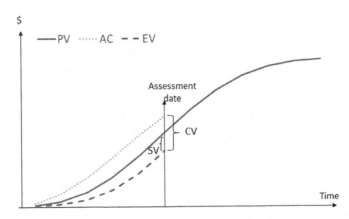

FIGURE 8.18 Blue curve: planned value (pv), yellow curve: actual cost (AC), and green curve: earned value (EV)

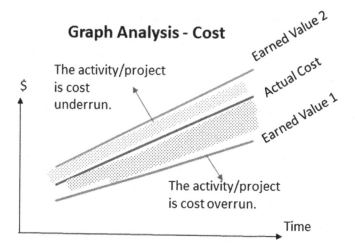

FIGURE 8.19 Earned value 2 curve is a cost underrun example, and earned value 1 curve is a cost overrun

The schedule is compared between the earned value curve and planned curve. Figure 8.20 shows the earned value curve 2 is above the planned curve, which means that the project or activity is ahead of schedule. On the contrary, like earned value curve 1, there is a delay because it is under the planned curve.

8.2.3 EVM FORECAST

The EVM forecast generates different scenarios from an optimistic to a pessimistic view. The calculation is based on three concepts, like below:

- BAC = Budget at completion. It is the project's budget or how much money is planned for execution.
- ETC = Estimate to completion. It is the money necessary to conclude the project. There are several methods to estimate it, and they are detailed below.
- EAC = Estimate at completion. It is the project's forecast or how much is forecasted at the end of the project. It can be higher, equal, or lower than the budget at completion. And it is the sum of estimate to completion plus the actual cost, like Equation 8.6.

$$\text{Estimate at completion (EAC)} = \text{AC} + \text{ETC} \qquad \text{(Equation 8.6)}$$

Figure 8.21 summarizes the three concepts, and in this scenario, the EAC is higher than BAC. It means that the project has a cost overrun, or more money than planned is necessary for completion.

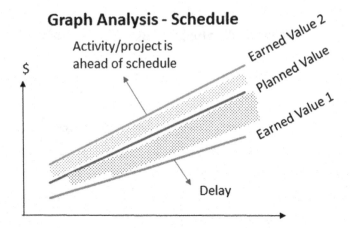

Graph Analysis - Schedule

FIGURE 8.20   Earned value 1 curve shows a delay, and earned value curve 2 means that the project/activity is ahead of schedule

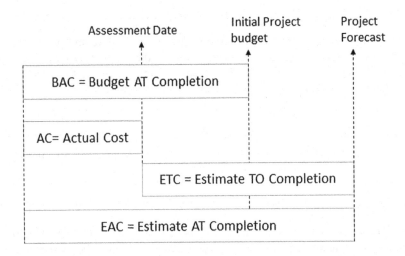

FIGURE 8.21   Budget at completion (bac), estimate to completion (etc), and estimate at completion (EAC)

There are four methods to calculate the estimate to completion (ETC) from a cost perspective.

Optimistic Scenario
It is assumed that the performance from the assessment date to the completion will be like the planning. It is estimated through Equation 8.7. Hence, there is no correlation with the project's actuals results.

$$\text{Estimate to completion (ETC)} = BAC - EV \qquad \text{(Equation 8.7)}$$

Realistic Scenario

The assumption is that performance until the project's end is similar to execution up-to-date, estimated through Equation 8.8.

$$\text{Estimate TO Completion } (ETC) = \frac{(BAC - EV)}{CPI} \quad \text{(Equation 8.8)}$$

Pessimistic Scenario

Cost and schedule performance will be similar to what has been done. It is calculated through Equation 8.9.

$$\text{Estimate TO Completion } (ETC) = \frac{(BAC - EV)}{(CPI \times SPI)} \quad \text{(Equation 8.9)}$$

Finally, a new estimate is recommended to define the ETC when the scope is modified, for example.

The planning adoption as a forecast is recommended for earlier project stages when the completion is below 10%, because the EVM forecast method scan promote distortions. Also, the manual method, when an expert defines the forecast, is recommended when the project completion is above 90% for the same reason. Figure 8.22 summarizes these recommendations.

Variance

After the EVM forecast calculation is possible to define the variance between the budget and forecast by the variance at completion through Equation 8.10.

$$\text{Variance at completion (VAC)} = BAC - EAC \quad \text{(Equation 8.10)}$$

Schedule Forecast

Also, the schedule at completion (SAC) can be estimated by Equation 8.11. It defines the schedule forecast through the up-to-date schedule performance divided by SPI.

$$\text{Schedule AT Completion } (SAC) = \frac{\text{Schedule}}{SPI} \quad \text{(Equation 8.11)}$$

8.2.4   EVM Forecast – Example 5

Estimate the forecast by the optimistic, pessimistic, and realistic scenarios after the ninth execution month using the process pilot plant data from Example 4. Also, calculate the variance at completion (VAC) and schedule at completion (SAC) considering the baseline in Figure 8.23.

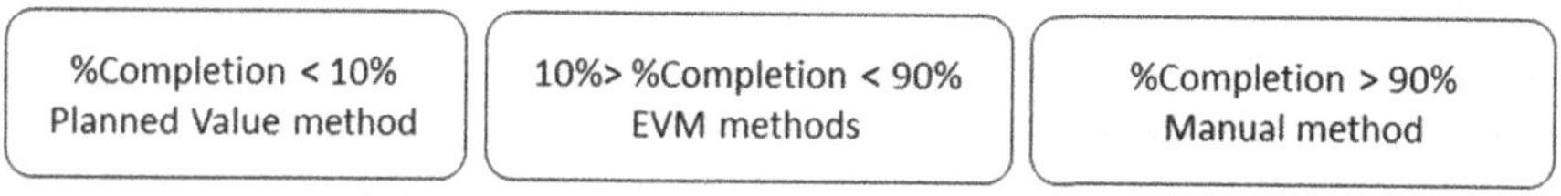

%Completion < 10% Planned Value method	10%> %Completion < 90% EVM methods	%Completion > 90% Manual method

FIGURE 8.22   Forecast method according to % completion

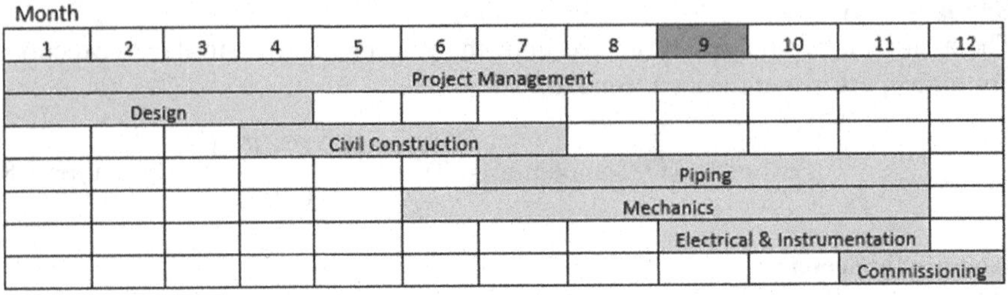

Month

1	2	3	4	5	6	7	8	9	10	11	12
				Project Management							
	Design										
			Civil Construction								
					Piping						
					Mechanics						
						Electrical & Instrumentation					
									Commissioning		

FIGURE 8.23   Original schedule (Baseline) – Example 5

TABLE 8.13
Optimistic Scenario Results

Discipline	Total planned (BAC) $ (year)	EV $	AC $	ETC $	EAC $	VAC $
Project management	1,225,000	918,750	930,500	306,250	1,236,750	(11,750)
Design	1,800,000	1,800,000	1,900,000	-	1,900,000	(100,000)
Civil construction	4,100,000	3,895,000	4,090,100	205,000	4,295,100	(195,100)
Piping	7,000,000	2,800,000	3,350,000	4,200,000	7,550,000	(550,000)
Mechanics	9,800,000	4,410,000	4,500,300	5,390,000	9,890,300	(90,300)
Electrical and instrumentation	4,600,000	690,000	650,000	3,910,000	4,560,000	40,000
Commissioning	2,500,000	-	-	2,500,000	2,500,000	-
Total	31,025,000	14,513,750	15,420,900	16,511,250	31,932,150	(907,150)

Optimistic Scenario

Table 8.13 shows the data from Example 4. Also, there is a summary of the project's ETC, EAC, and VAC for discipline.

Civil construction and piping are detailed below through Equations 8.6, 8.7, and 8.10 to exemplify the calculation.

Construction Civil – Optimistic Scenario

Estimate to completion (ETC) = BAC − EV = 4,100,000 − 3,895,000 = $ 205,000

Estimate at completion (EAC) = ETC + AC = 205,000 + 4,090,100 = $ 4,295,100

Variance at completion (VAC) = BAC − EAC = 4,100,000 − 4,295,100 = $ − 195,100

Piping – Optimistic Scenario

Estimate to completion (ETC) = BAC − EV = 7,000,000 − 2,800,000 = $ 4,200,000

Estimate at completion (EAC) = ETC + AC = 4,200,000 + 3,350,000 = $ 7,550,000

Variance at completion (VAC) = BAC − EAC = 7,000,000 − 7,550,000 = $ − 550,000

The civil construction variation, $ − 195,100 or − 5% (− 195,000/4,100,000), is considerable because it is the second highest variation. However, the discipline is near being finished, with 95% completion, so the result may become possible. The realistic scenario can confirm this statement.

Piping shows the highest variation, $ 550,000 or − 8% (− 550,000/4,200,000), which means a mitigation plan should be implemented.

Project variation, $ − 907,150 or − 3% (− 907,150/31,025,000), shows that the overrun cost will probably be inevitable unless a recovery plan is implemented within the last 3 months. After all, this is the optimistic scenario or the best scenario available.

Realistic Scenario

Table 8.14 shows the realistic scenario results. Civil construction and piping are detailed below through Equations 8.6, 8.8, and 8.10.

Construction Civil – Realistic Scenario

Estimate to completion (ETC) = (BAC − EV)/CPI = (4,100,000 − 3,895,000)/ 0.95 = $ 215,268

Estimate at completion (EAC) = ETC + AC = 215,268 + 4,090,100 = $ 4,305,368

Variance at completion (VAC) = BAC − EAC = 4,100,000 − 4,305,368 = $ − 205,368

TABLE 8.14
Realistic Scenario Results

Discipline	Total planned (BAC) $ (year)	EV	AC	CPI	ETC	EAC	VAC
Project management	1,225,000	918,750	930,500	0.99	310,167	1,240,667	(15,667)
Design	1,800,000	1,800,000	1,900,000	0.95	-	1,900,000	(100,000)
Civil construction	4,100,000	3,895,000	4,090,100	0.95	215,268	4,305,368	(205,368)
Tubulation	7,000,000	2,800,000	3,350,000	0.84	5,025,000	8,375,000	(1,375,000)
Mechanics	9,800,000	4,410,000	4,500,300	0.98	5,500,367	10,000,667	(200,667)
Electrical and instrumentation	4,600,000	690,000	650,000	1.06	3,683,333	4,333,333	266,667
Commissioning	2,500,000	-	-	N/A	2,500,000	2,500,000	-
Total	31,025,000	14,513,750	15,420,900	0.94	17,543,249	32,964,149	(1,939,149)

Piping – Realistic Scenario

Estimate to completion (ETC) = (BAC – EV) / CPI = (7,000,000 – 2,800,000)/0.84 = $ 5,025,000

Estimate at completion (EAC) = ETC + AC = 5,025,000 + 3,350,000 = $ 8,375,000

Variance at completion (VAC) = BAC – EAC = 7,000,000 – 7,550,000 = $ – 1,375,000

The civil construction forecast by the CPI should be analyzed carefully because the discipline completion is above 90%, as shown in Table 8.12 in Example 4. It was expected to finish in the seventh month, according to the baseline in Figure 8.23, but it still needs to be completed in the ninth month. Also, the variation is similar to the optimistic scenario, $ – 205,368 or – 5%, which could corroborate this estimation. Also, one attention point, which is not discussed here, is the impact between the disciplines or what is the delay promoted by civil construction in others and the critical path.

Piping becomes more dramatic because the variation is $ – 1,375,000 or – 20% showing a significant impact on the project. The reasons should be explained, justified, and actions implemented.

Realistic scenario shows the project has a $ – 1,939,149 variation or – 6% (– 1,939,149/31,025,000). It is double the optimistic scenario, and the red signal is on because the variation of two major disciplines, piping and mechanics, jumped from the optimistic scenario.

Pessimistic Scenario

Table 8.15 shows the pessimistic scenario results. Civil construction and piping are detailed below through Equations 8.6, 8.9, and 8.10.

TABLE 8.15
Pessimistic Scenario Results

Discipline	Total planned (BAC) $ (year)	EV	AC	CPI	SPI	ETC	EAC	VAC
Project management	1,225,000	918,750	930,500	0.99	1.00	310,167	1,240,667	(15,667)
Design	1,800,000	1,800,000	1,900,000	0.95	1.00	-	1,900,000	(100,000)
Civil construction	4,100,000	3,895,000	4,090,100	0.95	0.95	226,598	4,316,698	(216,698)
Tubulation	7,000,000	2,800,000	3,350,000	0.84	0.80	6,281,250	9,631,250	(2,631,250)
Mechanics	9,800,000	4,410,000	4,500,300	0.98	0.82	6,722,670	11,222,970	(1,422,970)
Electrical and instrumentation	4,600,000	690,000	650,000	1.06	1.00	3,683,333	4,333,333	266,667
Commissioning	2,500,000	-	-	N/A	N/A	2,500,000	2,500,000	-
Total	31,025,000	14,513,750	15,420,900	0.94	0.89	19,821,711	35,242,611	(4,217,611)

Construction Civil – Pessimistic Scenario

Estimate to completion (ETC) = (BAC – EV)/(CPI × SPI) = (4,100,000 – 3,895,000)/(0.95 × 0.95) = Estimate to completion (ETC) = $ 226,568
Estimate at completion (EAC) = ETC + AC = 226,568 + 4,090,100 = $ 4,316,698
Variance at completion (VAC) = BAC – EAC = 4,100,000 – 4,316,698 = $ – 216,698

Piping – Pessimistic Scenario

Estimate to completion (ETC) = (BAC – EV)/(CPI × SPI) = (7,000,000 – 2,800,000)/(0.84 × 0.80) =
Estimate to completion (ETC) = $ 6,281,250
Estimate at completion (EAC) = ETC + AC = 6,281,250 + 3,350,000 = $ 9,631,250
Variance at completion (VAC) = BAC – EAC = 7,000,00 – 9,631,250 = $ – 2,631,250

As expected, the two disciplines have a more dramatic scenario. It is because both indicators (CPI and SPI) are below one. However, piping has a huge variation, $ – 2,631,250 or – 38% (– 2,631,250/7,000,000) and it corresponds above 50% of the project's variation.

Figure 8.24 summarizes the range of possibilities for the project forecast from optimistic, realistic, and pessimistic scenarios. All show a cost overrun, but the level is from 3% to 14%. It shows that a mitigation plan could avoid a dramatic cost overrun where, for example, contingency is not enough to protect the company's cash flow.

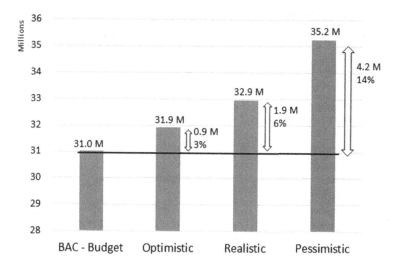

FIGURE 8.24 Project forecast according to each scenario

Schedule Forecast

The schedule forecast is estimated by Equation 8.11. For example, the project forecast is

Project schedule forecast = 12/SPI = 12/0.89 = 13.6 months

Table 8.16 shows the result by discipline. Piping and mechanical are the critical disciplines with a delay above 1 month planned.

8.2.5   EVM FORMULAS SUMMARY

Table 8.17 shows the EVM formulas summary.

TABLE 8.16
Schedule Forecast

Discipline	Schedule baseline (months)	Schedule forecast (months)
Project management	12	12.0
Design	4	4.0
Civil construction	4	4.2
Piping	5	6.3
Mechanical	6	7.3
Electrical and instrumentation	3	3.0
Commissioning	2	2.0
Total	12	13.6

TABLE 8.17
EVM Formulas Summary

Description	Abbreviation	Formula
Planned value	PV	PV
Actual cost	AC	AC
Earned value	EV	EV = % complete × budget for that account
Cost variance	CV	CV = EV − AC
Schedule variance	SV	SV = EV − PV
Cost performance index	CPI	CPI = EV/AC
Schedule peformance index	SPI	SPI = EV/PV
Budget at completion	BAC	BAC
Estimate to completion	ETC	According each scenario:
Optimistic scenario		ETC = (BAC − EV)
Realistic scenario		ETC = (BAC − EV)/CPI
Pessimistic scenario		ETC = (BAC − EV)/(CPI × SPI)
Estimate at completion	EAC	EAC = AC + ETC
Variance at completion	VAC	VAC = BAC − EAC
Schedule at completion	SAC	SAC = Schedule/SPI

8.2.6 EVM Challenges

Earned value management (EVM) is one of the most powerful project control techniques. However, it has challenges, as others methods as well. If scope changes occur and they are not analyzed and approved quickly, it can impact the analyses, and results will not reflect reality. The expected sequence is:

Scope Change ➜ Analyses ➜ Approval ➜ Baseline review ➜ EVM results considering the actual scope.

Unfortunately, if the data is *not* manipulated with transparency and ethics, the performance is good in the slides but not in practice. In addition to this attitude being unethical, it is not sustainable in the long term. And after discovery, it must be dealt with rigorously.

One key point is the schedule and cost control system integration, as shown in Figure 8.25, whereas the top image shows that the work breakdown for the cost and schedule do not match, but the down image shows the correct alignment between cost and cost schedule.

They must have the same EVM level of control. Also, if different software controls cost and time, they must be configured to ensure integration.

Finally, the measuring work progress defines the % of completion used in the earned value calculation. Hence, the correct selection of the method is crucial, as well as the application. The possible techiniques are:

- Supervision opinion
- Units completed
- Start/finish

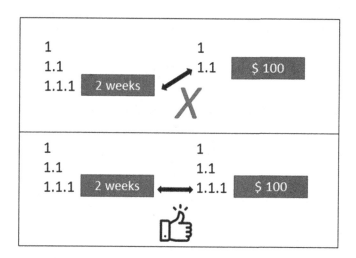

FIGURE 8.25   EVM level should be respected in cost and schedule control level and icons made by Freepick from www.flaticon.com

- Incremental milestone
- Cost ratio
- Weighted or equivalent units

8.3   S CURVE

The curve's shape names the method because it remembers the S format. According to AACE International, 2022, S Curve is *"in the context of project control, a cumulative distribution of costs, labour hours, progress, or other quantities plotted against time."*[5]

Figure 8.26 shows an S curve example.

Equation 8.12 shows the S curve equation:

$$\%\text{Completion}(n) = 1 - \left[1 - \left[\frac{n}{N}\right]^{\log I}\right]^{S}$$ (Equation 8.12)

Where:

n = Current period

N = Duration

I = Inflection point. It is the point when the curve changes its concavity.

S = Shape coefficient. It depends on the activity or project type.

There are typical S curves, such as 30%, 40%, 50%, and 60% S curves.

30% S Curve ➜ It means that the project (or activity) has 50% of completion when 30% duration is achieved.

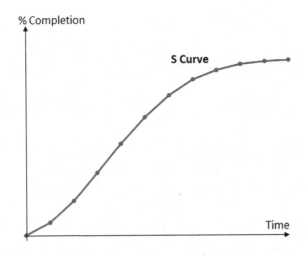

FIGURE 8.26   S curve

40% S Curve ➜ It means that the project (or activity) has 50% of completion when 40% duration is achieved.

50% S Curve ➜ It means that the project (or activity) has 50% of completion when 50% duration is achieved.

60% S Curve ➜ It means that the project (or activity) has 50% of completion when 60% duration is achieved.

Figure 8.27 shows the four curves, and Table 8.18 shows the I, S, and values adopted in Equation 8.12 to generate them.

Graph analysis is a crucial benefit of the S curves. Figure 8.28 shows two scenarios; actual 1 (dashed curve) is above the planned curve, which means there is a

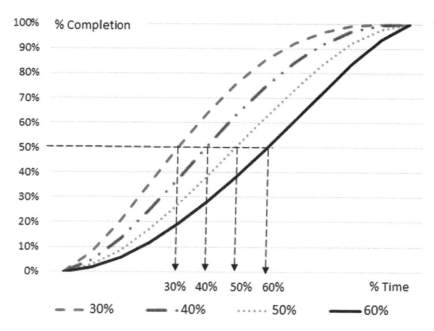

FIGURE 8.27   Typical S curves of 30%, 40%, 50%, and 60%.

TABLE 8.18
I, S, and Values Adopted in Equation 8.12
to Generate S Curves in Figure 8.27

	30%	40%	50%	60%
I	30	40	50	60
S	3.2	2.5	1.9	1.4
N	12	12	12	12

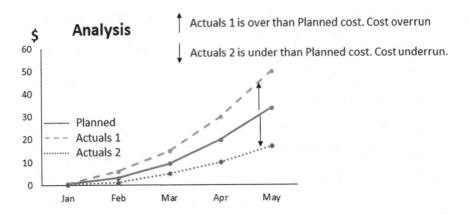

FIGURE 8.28   S curve – vertical analysis

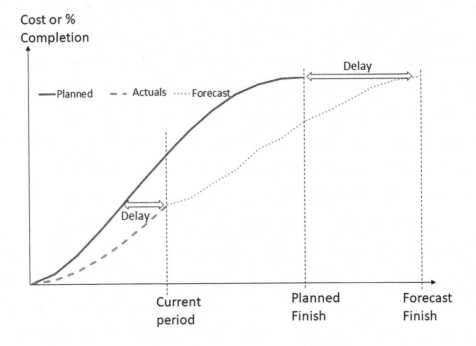

FIGURE 8.29   S curve – horizontal analysis

cost overrun. On the other hand, actual 2 (dotted curve) is under the planned curve showing that performance is a cost underrun. Hence, it promotes a quick and visual analysis of the project's performance.

In addition to the vertical analysis, the horizontal analysis helps check delays. Figure 8.29 shows that horizontal variation highlights a current delay (small arrow), and the forecast has a considerable delay (longest narrow).

8.3.1 EVM and'S Curve – Example 6

Project Blue Sky has 12 months and a $ 1,000 baseline. After 6 months, the actual cost, earned value, and forecast are shown in Figure 8.30. The example aims to show the S curve and EVM analysis. Estimate SV, CV, VAC, ETC, ETC, EAC, and SAC.

The graph shows the project has a cost overrun in the current period because the actual curve (dashed curve) is over the earned value curve (solid curve). Also, there is a delay because the earned value curve is under the planned curve (dashed and dotted curve).

The scenario becomes worse when the forecast is analyzed because, visibly, the horizontal and vertical variations increase significantly. It means a considerable delay and a cost overrun. The calculations below can confirm the analysis, and Figure 8.31 illustrates it.

Calculations based on formulas from Table 8.17:

$CV = 383 - 1,151 = -768$
$SV = 383 - 632 = -249$
$CPI = 383/1,151 = 0.33$
$SPI = 383/632 = 0.61$
$BAC = 1,000$
$ETC = (1,000 - 383)/0.33 = 1,870$ – Using the realistic scenario
$EAC = ETC + AC = 1,870 + 1,151 = 3,021$
$VAC = 1,000 - 3,021 = -2,021$
$SAC = 12/0.61 = 19.7$ months or ~20 months

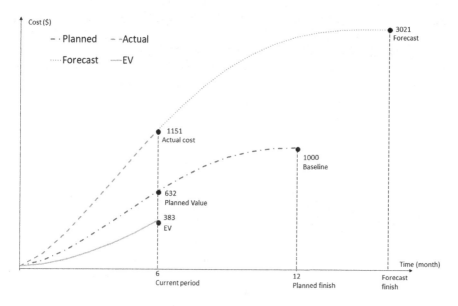

FIGURE 8.30 Planned, actual cost, earned value, and forecast curves after 6 months – project blue sky – Example 6

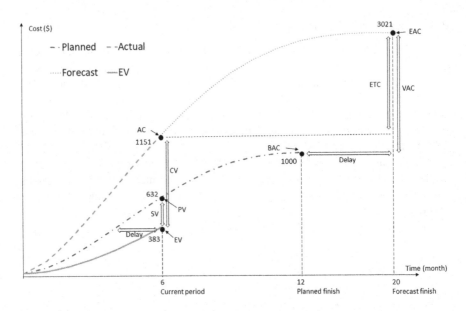

FIGURE 8.31   CV, SV, ETC, EAC, VAC, and delays – project blue sky – Example 6

8.4   RUNDOWN CURVE

The rundown curve is a method to control activities, making a comparison with what was planned. It represents in a decreasing way the activity production like, for example, manufacture or assembly pipes and the laying of cables. It can also be used to control contingency.

Figure 8.32 shows a rundown curve example. It is a piping assembly activity whose duration is 12 months, and the planned quantity is 1,200 tons. The production or reduction is highlighted between the fourth and fifth months, which is 100 tons.

The analysis from the rundown curve is demonstrated in Figure 8.33. The activity could be delayed if the performance is above the planned curve. In this case, it is necessary to define an action plan to change this trend. However, if the execution curve is under the planned curve, the activity performs better than the plan. It could be maintained, or some resources could be managed for other activities.

8.4.1   RUNDOWN CURVE EXAMPLE – EXAMPLE 7

The piping activity mentioned in Figure 8.32 is analyzed after 7 months of execution, like Figure 8.34. It is possible to observe that until May, the activity's performance is worse than planned. In addition, in May and June, there is no activity. Consequently, a line parallel to the x-axis is shown. Finally, the forecast curve shows a delay of 3 months from what was planned. It is necessary to include more resources or adopt techniques that increase productivity to change this scenario.

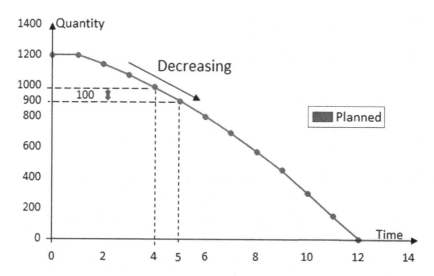

FIGURE 8.32   Rundown curve example

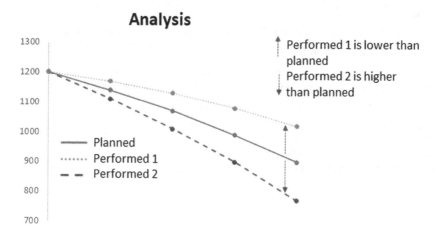

FIGURE 8.33   Rundown curve analysis

Table 8.19 shows the planned, performed and forecast values illustrated in Figure 8.34, whereas the delay of 3 months and the performed values under the planned value indicate worse performance.

8.5   INDICATORS

Indicators work as traffic lights. For example, green means the result is satisfactory, yellow requires attention, and red is an out-of-goal.

Indicators promote clear communication, strategy alignment, progress measuring, and monitoring.

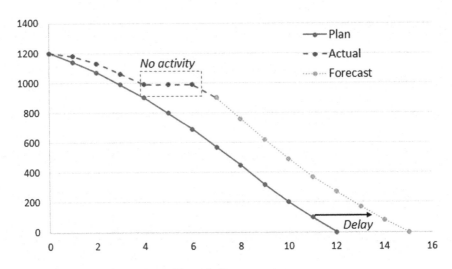

FIGURE 8.34　Rundown curve – Example 7

TABLE 8.19

Planned, Production Planned, Performed, Production Performed and Forecast Values – Example 7

Time	Planned (ton)	Planned production (ton)	Performed (ton)	Performed production (ton)	Forecast (ton)
Jan	1,200	60	1,200	20	
Feb	1,140	70	1,180	50	
Mar	1,070	80	1,130	70	
Apr	990	90	1,060	70	
May	900	100	990	0	
Jun	800	110	990	0	
Jul	690	120	990	90	
Aug	570	120	900		
Sep	450	130			760
Oct	320	120			620
Nov	200	100			490
Dec	100	100			370
Jan	0	0			270
Feb					170
Mar					80
Apr					0

The calculation should be standardized and easily understood as an absolute value (e.g., target) or a rate. For example, safety indicators should be understandable for all team members because it helps to assimilate and implement the desired values reducing risks of incidents and accidents.

Complex or many indicators could reduce their effectiveness because they could not be fully understood. Hence, it can impact their control.

Indicators are becoming more popular, making it possible to navigate different views thanks to business intelligence (BI) dashboards.

As in any control, manipulations for the indicators estimation, when the data is modified to ensure the wished result, is a risk that should be mitigated because it can affect the company result.

One famous method is the balanced scorecard (BSC), a strategic planning and management system. A scorecard is a valuable tool because it works as an airplane cockpit and the presentation of both financial and operational measures.

The quantity and type of indicators can vary by industry or department. If possible, an innovation indicator should be considered. Also, a BSC should not have many indicators. And it can have an overall score by Equation 8.13:

$$\text{Overall } (\%) = \sum f_i \, (P_i' \, / \, P_i) \, / \sum f_i \qquad \text{(Equation 8.13)}$$

f = Factor
i = Number of indicator
P' = Performed
P = Planned

Furthermore, the BSC can have a monthly or annual vision, so the periodicity must be defined.

8.5.1   SCORECARD – EXAMPLE 8

The ABC Company defined each manager's scorecard based on five top indicators: CAPEX, OPEX, safety, revenue, and customer satisfaction. Each indicator has a target and an overall score for which the goal is equal to or above 100%. Also, the analysis should be done monthly.

Figure 8.35 shows each indicator's plan: actual, result, goal, and status for the Subsea installations department in March 2022.

The result for each indicator is a simple rate actual by the plan. For example:

$$\text{CAPEX} = 45/50 = 90\%$$

The factors to estimate the overall are shown below and calculated by Equation 8.13:

Factor
CAPEX   2
OPEX   1
Safety   2
Revenue   3
Customer satisfaction 2

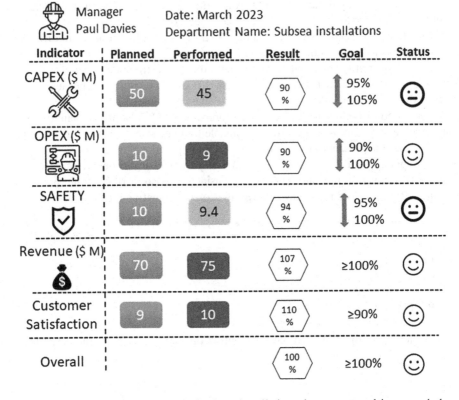

FIGURE 8.35 Balance scorecard of subsea installations department and icons made by Freepick from www.flaticon.com

$$\text{Overall } (\%) = \frac{2 \times (45/50) + 1 \times (9/10) + 2 \times (9.4/10) + 3 \times (75/70) + 2 \times (10/9)}{2 + 1 + 2 + 3 + 2}$$

$$= \frac{10.2}{10} = 100.2\%$$

Although CAPEX and safety indicators are off target, the overall result is satisfactory.

NOTES

1-5. Reprinted with permission from AACE International. Check the website for the latest versions (https://web.aacei.org/resources/cost-engineering-terminology).

BIBLIOGRAPHY

AACE International, 2022. *Recommended Practice 10S-90 Cost Engineering Terminology.* Available at: https://web.aacei.org/resources/cost-engineering-terminology. [Accessed: 8 January 2023].

Drury, C., 2015. *Management and Cost Accounting – Student Manual*, 9th Edition. Cengage Learning.

GAO, 2020. *Cost Estimating and Assessment Guide*. Available at: https://www.gao.gov/products/gao-20-195g. [Accessed: 15 December 2022].

Hastak, M., 2015. *Skills & Knowledge of Cost Engineering*, 6th Edition. AACE International.

Herold, T., 2014. *Financial Terms Dictionary*. Evolving Wealth, LLC.

Kaplan, R., Norton, D., 1992. *The Balanced Scorecard—Measures that Drive Performance*. Available at: https://hbr.org/1992/01/the-balanced-scorecard-measures-that-drive-performance-2. [Accessed: 15 April 2023].

Oxford, 2016. *Dictionary of Business and Management*, 6th Edition. Oxford University Press.

RICS, 2015. *Cost Reporting*, 1st Edition. RICS Professional Guidance, UK. Available at: https://www.rics.org/profession-standards/rics-standards-and-guidance/sector-standards/construction-standards/black-book/cost-reporting. [Accessed: 25 April 2023].

Tennent, J., 2013. *Guide to Financial Management*, 2nd Edition. The Economist in Association with Profile Books Ltd.

Whiteley, J., 2004. *Financial Management*. Palgrave Master Series.

9 Additional Topics

The chapter covers topics that could be used during cost estimation, cost control, and economic feasibility study. They are listed below:

- Benchmarking
- Depreciation
- Inflation and escalation
- Sensitivity analysis
- Value engineering
- Contract cost updates

9.1 BENCHMARKING

Benchmarking is a process used by many industries aiming to improve performance through benchmark comparisons. The comparison could be with internal or external benchmarking. Internal benchmarking is based on the best projects or processes inside the company, and external is out of the organization, from other companies.

It has many applications, such as supporting decision-making, project planning, and cost and schedule estimates analysis and validation. For example, the cost validation process could use benchmarks to ensure that the result is similar to the industry benchmarks. An outlier result could mean an error during the cost estimate process or assumptions are super optimistic/pessimistic.

The benchmarking elaboration starts with data acquisition from previous or similar projects, as shown in Figure 9.1. Then, the data should be normalized as discussed in Section 3.5, and the assessment could be done through statistics measures such as mean, median, minimum, maximum, or regression analysis.

Finally, the benchmark should be tested and documented, allowing traceability.

Figure 9.2 shows a benchmarking application example. Two metrics represented by continuous lines and their industry benchmark (diamond) evaluate two similar projects. Also, the dotted lines show the confidence interval. It means that a result within these ranges is considered acceptable.

Both projects have good results for the scheduling metric, but project A (rectangle) has an excellent result under the industry benchmark and between the confidence intervals. However, project B (hexagon) could be better because it is over the benchmark. For the cost metric, project A has opportunities to improve compared to the industry benchmark. Project B needs to be reevaluated because the result is not satisfactory.

DOI: 10.1201/9781003402725-9

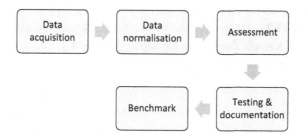

FIGURE 9.1   Benchmarking process

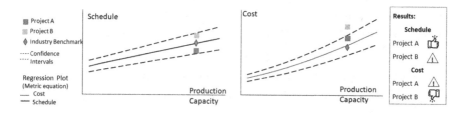

FIGURE 9.2   Benchmarking example and icons made by Freepick from www.flaticon.com

9.1.1   BENCHMARKING ESTIMATION – EXAMPLE 2

The following example is based on internal benchmarking estimation for a vessel elaborated by the company Blue Sea. The assumption is all vessels are the same material and specifications, except the volume, and ten acquisitions per size are enough to calculate a benchmark (in real situations, improving the number to 50 is recommended). Also, they are bought from different suppliers but in the same country. Table 9.1 shows the vessel cost per size and acquisition date.

The normalization is done by Equation 3.5, and the hypothetical economic index is shown in Table 9.2. The date to make the normalization is September 2023.

The cost update for vessel one is shown below to exemplify the normalization:

Vessel 1 cost updated = 140,000 × 233/200 = $ 163,100

Figure 9.3 shows the result for all vessels after the normalization. Finally, company Blue Sea makes two assumptions: the benchmark is the percentile 20, and the acceptable limit range for future acquisitions is the percentile 70.

Benchmark 20 m³ = $ 165,152 and limit range = $ 183,361
Benchmark 40 m³ = $ 351,749 and limit range = $ 374,797

The size 20 m³ results show one vessel is under the benchmark, vessel 03, and three (vessels 5, 9, and 10) exceed the limit range. The size 40 m³ has two vessels under the benchmarking (vessels 11 and 12), and three exceed the limit (vessels 13, 15, and 20). It should support future acquisition and negotiation processes.

TABLE 9.1

List of Vessels Bought by Company Blue Sea –
Benchmarking Example 2

Description	Value ($)	Capacity (m³)	Acquisition date
Vessel 01	140,000	20	Jan-22
Vessel 02	150,000	20	Mar-22
Vessel 03	125,000	20	Jan-22
Vessel 04	155,000	20	Oct-22
Vessel 05	190,000	20	Jan-23
Vessel 06	170,000	20	Dec-22
Vessel 07	160,000	20	Feb-23
Vessel 08	180,000	20	Jul-23
Vessel 09	165,000	20	May-22
Vessel 10	190,000	20	Aug-23
Vessel 11	320,000	40	Sep-22
Vessel 12	330,000	40	Jan-23
Vessel 13	330,000	40	Feb-22
Vessel 14	350,000	40	Mar-23
Vessel 15	340,000	40	Jun-22
Vessel 16	350,000	40	Nov-22
Vessel 17	360,000	40	May-23
Vessel 18	320,000	40	Jan-22
Vessel 19	330,000	40	Oct-22
Vessel 20	370,000	40	Jun-23

The benchmark decision could be only based on a statistic measure, like in this example, or a manager/specialist decision considering their expertise.

9.1.2 BENCHMARKING – EXERCISE 1

Company ABC wants to analyze three projects. Considering the graph in Figure 9.4 and the data below, what is your recommendation per project? Can they go ahead, or should they be reviewed? The solid line represents the industry benchmark equation which correlates the factory capacity (m³/ton) and CAPEX cost ($ millions). For example, the industry benchmark for 500 m³/ton capacity is $ 56 million or the dot in the graph.

Project A → CAPEX $ 25 million and capacity 250 m³/day
Project B → CAPEX $ 40 million and capacity 350 m³/day
Project C → CAPEX $ 60 million and capacity 400 m³/day

The black dash lines show the upper and lower limits defined by Company ABC.

TABLE 9.2

Vessel Economic Index – Example 2

Period	Index
Jan-22	200
Feb-22	202
Mar-22	204
Apr-22	206
May-22	207
Jun-22	207
Jul-22	209
Aug-22	213
Sep-22	216
Oct-22	218
Nov-22	218
Dec-22	220
Jan-23	221
Feb-23	223
Mar-23	223
Apr-23	225
May-23	227
Jun-23	229
Jul-23	230
Aug-23	232
Sep-23	233

9.2 DEPRECIATION METHODS

Depreciation is important to encourage new investments. The investment depreciation allows the firm to reduce its income by a yearly proportion with a depreciation write-off until the asset is fully depreciated.

AACE International, 2022, defines depreciation as *"Decline in value of a capitalized asset. A form of capital recovery applicable to a property with a life span of more than one year, in which an appropriate portion of the asset's value is periodically charged to current operations."*[1] In addition, depletion is analogous to depreciation but for natural resources (e.g., coal, oil, and timber in forests).

There are several methods to calculate depreciation, such as straight-line (SL), double-declining balance (DDB), sum-of-years digits (SOYD), and units of production (UOP).

9.2.1 STRAIGHT-LINE (SL) METHOD

The method is based on a linear depreciation along the asset life, excluding the salvage value. Equation 9.1 shows the SL method equation.

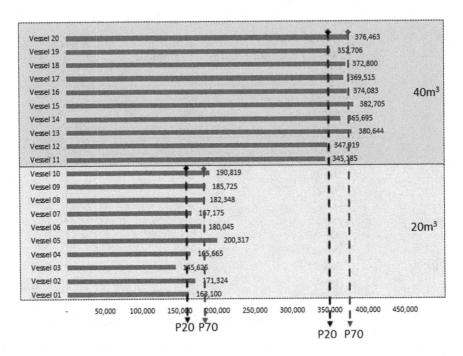

FIGURE 9.3   Vessel cost after normalization, benchmark (p20), and range limit (p70) per size – Example 2

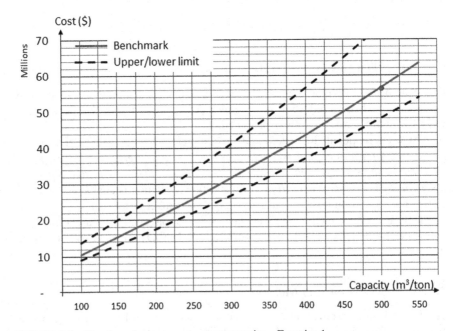

FIGURE 9.4   Benchmark chart, cost versus capacity – Exercise 1

$$SL_d = \frac{C - SV}{N}$$ (Equation 9.1)

Where:

SL$_d$ = Depreciation charge
C = Asset original cost
N = Asset depreciable life (years)
SV = Salvage value. It means the estimated book value of an asset after depreciation is complete. It is how much the organization intends to receive for the asset at the end of its useful life.

Suppose that Company ABC has an asset costing $ 10,000 and the asset life is 5 years. What is the depreciation charge through the stratight-line method considering a $ 2,000 salvage value?

Through Equation 9.1

SL$_d$ = (10,000 − 2,000)/5 = $ 1,600

Figure 9.5 shows that depreciation is the same amount per year until achieving the salvage value after 5 years.

9.2.2 DOUBLE-DECLINING BALANCE (DDB) METHOD

Double-declining balance (DDB) is an accelerated method promoting fast depreciation in the early years. Equation 9.2 shows how the depreciation charge is estimated.

$$DDB_d = Time \times BV$$ (Equation 9.2)

Where

DDB$_d$ = Depreciation charge

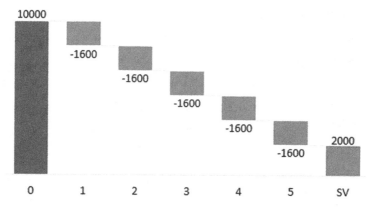

FIGURE 9.5 Straight-line method example

Time = 2/asset depreciable life (years)

BV = Book value at a given year. It means the asset value is subtracted by accumulated depreciation.

Suppose Company ABC has an asset costing $ 10,000 and 5 years of asset life. What is the depreciation charge through the DDB method considering a $ 2,000 salvage value?

Equation 9.2 calculates each year's depreciation as listed in Table 9.3.

Figure 9.6 shows the depreciation charge per year, and year 5 has zero depreciation because the asset was totally depreciated until year 4.

9.2.3 DDB AND SL – EXAMPLE 5

Company ABC can depreciate an asset of $ 100,000 through DDB or SL methods. However, the asset depreciable life is different, 10 years for SL and 15 years for the

TABLE 9.3
DDB Example 4

	Results	
Year	Book value ($)	Depreciation charge
1	10,000	2/5 × 10,000 = 4,000
2	10,000 – 4,000 = 6,000	2/5 × 6,000 = 2,400
3	6,000 – 2,400 = 3,600	2/5 × 3,600 = 1,440
4	3,600 – 1,440 = 2,160	2/5 × 2,160 = 160*
5	2,000	0

*The depreciation charge is $ 160 in year 4 because it could not be lower than the salvage value.

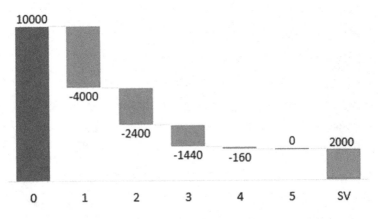

FIGURE 9.6   Double-declining balance method example

DDB technique. Company ABC wants the technique with higher depreciation during the first 5 years. What method should be selected considering a salvage value of $ 15,000?

SL Method

$$\text{Depreciation Charge} = \frac{100,000 - 15,000}{10}$$

$$= \$\,8,500$$

Depreciation up year 5 = 5 × 8,500 = **$ 42,500**

DDB Method

The DDB method result per year is listed in Table 9.4.

Depreciation up year 5 = 13,333 + 11,556 + 10,015 + 8,680 + 7,522 = **$ 51,105**

The DDB, as shown in Figure 9.8, promotes a higher depreciation than the SL method, despite 15 years of depreciable asset life. It confirms that it is suitable for faster amortization in the first years.

9.2.4 SUM-OF-YEARS DIGITS (SOYD)

Sum-of-years digits (SOYD) is a fasted depreciation method that promotes higher depreciation during the first years. Equation 9.3 shows how to estimate the depreciation charge per year.

TABLE 9.4
DDB Example 5

	Results	
Year	Book value ($)	Depreciation charge
1	100,000	13,333
2	86,667	11,556
3	75,111	10,015
4	65,096	8,680
5	56,417	7,522
6	48,895	6,519
7	42,375	5,650
8	36,725	4,897
9	31,829	4,244
10	27,585	3,678
11	23,907	3,188
12	20,719	2,763
13	17,957	2,394
14	15,562	562
15	15,000	-

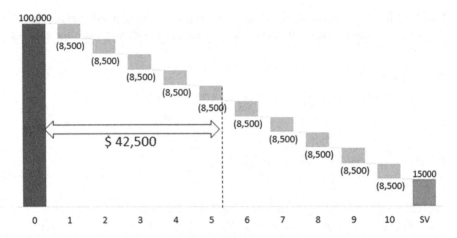

FIGURE 9.7 SL depreciation per year – Example 5

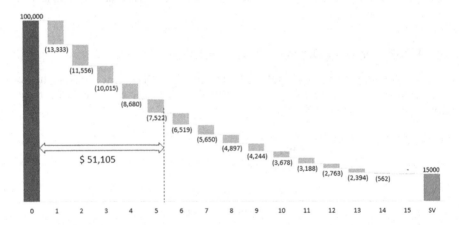

FIGURE 9.8 DDB depreciation per year – Example 5

$$SOYD_d = \frac{(C - SV) \times Y}{[\text{Sum-of-years}]}$$ (Equation 9.3)

Where
  $SOYD_d$ = Depreciation charge
  C = Asset original cost
  N = Asset depreciable life (years)
  Y = N + 1 – current year
  Sum-of-years = $(N^2 + N)/2$
  SV = Salvage value

Through the same data from previous methods, Company ABC wants to depreciate an asset of $ 10,000, life of 5 years, and SV 2,000.

TABLE 9.5
SOYD Depreciation Calculation

Results SOYD

Year	Y	Depreciation charge ($)
1	Y = 5 + 1 − 1 = 5	(10,000 − 2,000) × 5/15 = 2,667
2	Y = 5 + 1 − 2 = 4	(10,000 − 2,000) × 4/15 = 2,133
3	Y = 5 + 1 − 3 = 3	(10,000 − 2,000) × 3/15 = 1,600
4	Y = 5 + 1 − 4 = 2	(10,000 − 2,000) × 2/15 = 1,067
5	Y = 5 + 1 − 5 = 1	(10,000 − 2,000) × 1/15 = 533

The sum-of-years is $(5^2 + 5)/2 = 15$, and Table 9.5 shows the calculation per year.

Figure 9.9 shows the yearly depreciation through the SOYD method, and the higher depreciation occurs during the first years of the asset life.

The sequence below compares the faster-to-slow method. Then, the DDB is the fastest technique, and SL is the slowest.

$$DDB > SOYD > SL$$

9.2.5 Units of Production (UOP)

The UOP method is typically applied to assets used in the production line. The depreciation charge is estimated by Equation 9.4.

$$UOP_d = \frac{(C - SV) \times NUy}{n}$$ (Equation 9.4)

Where:
UOP_d = Depreciation charge
C = Asset original cost
NUy = Number of units produced in the year
n = Estimated total units of production during the asset's useful life
SV = Salvage value

Company ABC wants to depreciate an asset through the data in Table 9.6 by the UOP method.

The highest depreciation, the circle, occurs when the annual production is the biggest, as shown in the estimation below and Figure 9.10.

Year 1 ➔ Depreciation = (10,000 − 2,000)/5,000 × 1,000 = $ 1,600
Year 2 ➔ Depreciation = (10,000 − 2,000)/5,000 × 2,000 = $ 3,200
Year 3 ➔ Depreciation = (10,000 − 2,000)/5,000 × 500 = $ 1,600
Year 4 ➔ Depreciation = (10,000 − 2,000)/5,000 × 500 = $ 800
Year 5 ➔ Depreciation = (10,000 − 2,000)/5,000 × 500 = $ 800

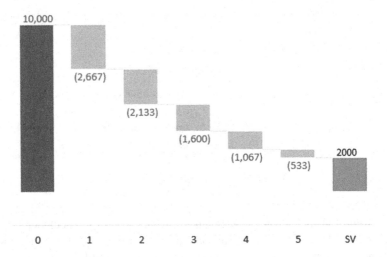

FIGURE 9.9   SOYD depreciation per year – Example 6

TABLE 9.6
UOP Depreciation Calculation

Input – UOP		
Description	**Value**	**Unit**
Book value – Initial	10,000	$
Salvage value	2,000	$
Asset depreciable life	5	years
Production per year		
Year 1	1,000	Units
Year 2	2,000	Units
Year 3	1,000	Units
Year 4	500	Units
Year 5	500	Units
Total (n)	5,000	Units

9.2.6 DEPRECIATION EXERCISE 2

Company Blue Sea has four assets that should be depreciated according to the data and methods listed below. Calculate the depreciation charge per year and asset. Also, what is the portfolio yearly depreciation, assuming all asset depreciation starts in the same year?

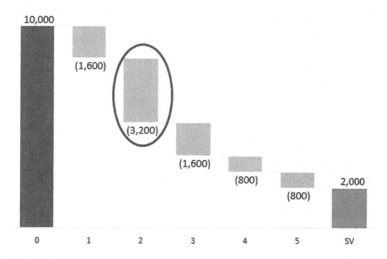

FIGURE 9.10 UOP depreciation per year – Example 7

Asset 1 → Green Building
Original cost – $ 10 million
Asset depreciable life – 20 years
Salvage value – $ 1 million
Depreciation method – Straight-line (SL)

Asset 2 → Blue Factory
Original cost – $ 25 million
Asset depreciable life – 25 years
Salvage value – $ 5 million
Depreciation method – Double-declining balance (DDB)

Asset 3 → White Road
Original cost – $ 20 million
Asset depreciable life – 25 years
Salvage value – $ 2 million
Depreciation method – Sum-of-years-digits (SOYD)

Asset 4 → Red Factory
Original cost – $ 30 million
Asset depreciable life – 20 years
Salvage value – $ 5 million
Depreciation method – Units of production (UOP)
Total production estimated – 20,000 units
Production estimated per year
Year 1 = 600 units
Year 2 = 800 units
Year 3 = 1,200 units
Year 4 = 1,500 units
Year 5 = 1,700 units

Year 6 = 2,000 units
Year 7 = 1,900 units
Year 8 = 1,700 units
Year 9 = 1,500 units
Year 10 = 1,200 units
Year 11 = 1,000 units
Year 12 = 800 units
Year 13 = 600 units
Year 14 = 500 units
Year 15 = 500 units
Year 16 = 500 units
Year 17 = 500 units
Year 18 = 500 units
Year 19 = 500 units
Year 20 = 500 units

9.3   INFLATION AND ESCALATION

Inflation and escalation have similarities, but they are different concepts. Escalation is related to a specific product, service, or commodity price rise due to factors such as inflation, engineering changes, environmental effects, and supply versus demand. Hence, inflation is an escalation component, but they are not necessarily equal.

AACE International, 2022, defines inflation as *"persistent increase in the level of consumer prices, or a persistent decline in the purchasing power of money, caused by an increase in available currency and credit beyond the proportion of available goods and services."*[2]

Related factors of both concepts are illustrated in Figure 9.11. Inflation is directly correlated with purchasing power. When inflation rises, the purchasing power

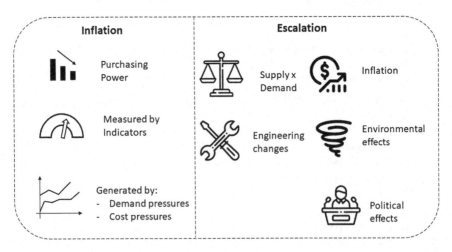

FIGURE 9.11 Escalation and inflation factors and icons made by Freepick from www.flaticon.com

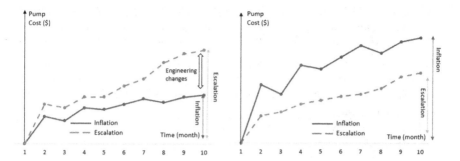

FIGURE 9.12   Hypothetical scenarios for a pump cost

decreases and vice versa. Inflation can be measured by indicators such as the consumer price index. Also, cost and demand pressures raise inflation.

Figure 9.11 (right) shows the effects that may promote escalation. The balance of supply and demand can affect the price of any product or service. Also, if there is a new requirement by engineering, it can increase the cost, for example, of new accessories to improve safety. Political effects are related to new laws and legislation increasing taxes or subsidies. An event of force majure, such as a hurricane or tornado, can dramatically affect the cost because it suddenly promotes a disbalance in supply versus demand.

Figure 9.12 (left) shows a hypothetical scenario where the pump cost increases because of the escalation (dashed line). Also, inflation (solid line) is one of the components of escalation, but engineering changes also contribute. The right figure shows a different result, whereas the escalation is lower than the inflation because the imbalance of supply versus demand promotes a cost reduction, for example.

The cost due to the escalation is calculated by applying the escalation % through the original cost, like below:

$$\text{Cost'} = \text{Original cost} \times (1 + \%\ of\ escalation) \qquad \text{(Equation 9.5)}$$

For example, if the escalation during the last 2 years is 7% and the pump original cost is $ 10,000, the updated cost is:

$$\text{Cost'} = 10,000 \times (1 + 0.07) = \$\ 10,700$$

The escalation could apply to each component of the cost estimate or a group, such as steel components, valves, pumps, etc. Or, it could apply to the top level, as illustrated in Section 7.7.

9.4   SENSITIVITY ANALYSIS

The technique is based on re-estimate key parameters to analyze the final result and support the decision-makers. The GAO, 2020, defines sensitivity analysis as *the recalculating the cost estimate with different quantitative values for selected inputs to compare the results with the original estimate. If a small change in the value of a factor yields a large change in the overall cost estimate, the results are considered sensitive to that factor.*

Figure 9.13 summarizes the process by identifying the cost drivers or parameters to be tested. Then, the estimated cost should be recalculated. Finally, the results are analyzed and documented.

The example below aims to analyze the cost impact of six factors in a cost estimate of $ 1 million. They are:

Factor 1 – Increase steel weight
Factor 2 – Improve the learning curve
Factor 3 – Subcontracting the welding service
Factor 4 – Increase testing and quality requirements
Factor 5 – Subcontracting the design
Factor 6 – Rework

After the parameters identification, the re-estimate is done for each assumption. Table 9.7 shows the original cost breakdown and the result of each re-estimate. Rework, factor 3, has the most increased effect, and subcontracting the design, factor 5, is the most significant reduction.

The increase in steel weight, factor 1, illustrates the process in Figure 9.14. The assumption is the steel quantity increases by 50%, and the re-estimate generates 9% over the original estimate because of the additional $ 60,000 material and $ 30,000 labor (direct + indirect).

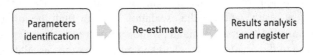

FIGURE 9.13   Sensitivity analysis sequence

TABLE 9.7
Sensitivity Analyses Example – Re-estimate Results

	Original cost ($)	Factor 1 ($)	Factor 2 ($)	Factor 3 ($)	Factor 4 ($)	Factor 5 ($)	Factor 6 ($)
Direct labor	300,000	324,000	291,000	285,000	345,000	300,000	360,000
Indirect labor	150,000	156,000	150,000	145,500	163,500	105,000	165,000
Material	300,000	360,000	300,000	300,000	345,000	300,000	345,000
Construction equipment	40,000	40,000	40,000	40,000	40,000	40,000	40,000
Overhead	50,000	50,000	50,000	50,000	50,000	50,000	50,000
Subcontract	80,000	80,000	80,000	84,000	80,000	92,000	80,000
Office and others	80,000	80,000	80,000	80,000	80,000	80,000	80,000
Total	**1,000,000**	**1,090,000**	**991,000**	**984,500**	**1,103,500**	**967,000**	**1,120,000**

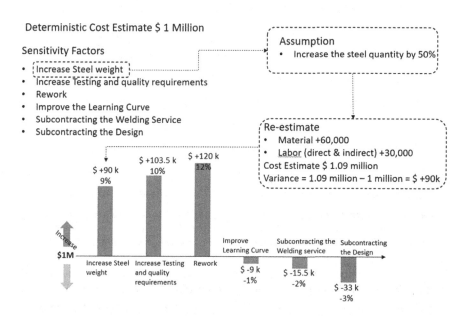

FIGURE 9.14   Sensitivity analysis example

Hence, sensitivity analysis provides a powerful vision of cost drivers and highlights the attention points along the planning and execution phases. For example, avoiding or mitigating rework is crucial for this scope.

9.4.1 SENSITIVITY ANALYSIS EXERCISE 3

Considering the cost estimate calculated in the Section 4.11, execute a sensitivity analysis considering the four factors below:

Factor 1 – Tower cost is revised to $220,000
Factor 2 – The benefits are revised:
   Direct fringe benefits = 48% and
   Indirect fringe benefits = 40%
Factor 3 – The lift services are subcontracted, costing $ 30,000 monthly. Hence, no lift team in the direct cost and no rent truck and cranes cost should be adopted.
Factor 4 – Interest rate increases to 5% and tax to 25%

What is the factor that generates the highest cost impact?

9.5   VALUE ENGINEERING (VE)

Value engineering (VE) is a technique which aims to identify and measures solutions promoting cost reduction or improving performance. AACE International, 2022, defines it as

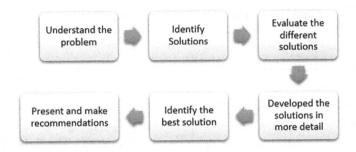

FIGURE 9.15 Value engineering sequence

> *a practice function targeted at the design itself, which has as its objective the develop-
> ment of design of a facility or item that will yield least life-cycle costs or provide great-
> est value while satisfying all performance and other criteria established for it.*[3]

Figure 9.15 shows the steps to implement the technique, starting with understanding the problem and identifying possible solutions. After that, each possible answer
should be analyzed and developed, providing scenarios to support the decision-makers. Then, the best solution is placed, and the final report makes the recommendations for the project.

The example is based on an O&G process plant analyzing possible solutions for
a flare. It works as safety equipment, which ensures the excess gas combustion that
cannot be recycled or recovered and reduces the emission of pollutants.

The idea is to analyze the possible solutions to the flare structure. Three options
are explored: Derrick, Guyed, or self-supported, as shown in Figure 9.16. Although
the decision should consider many aspects, such as construction, quality, maintenance, and operation, only the cost perspective is analyzed.

The Table 9.8 shows the CAPEX, OPEX, and operation life per structure type.
The Derrick option has the best result despite the higher CAPEX.

9.5.1 VALUE ENGINEERING – EXERCISE 4

Company ABC wants to analyze two possible solutions for a process plant in the
refinery through the VE technique. Option 1 is based on the unique vessel, and
option 2 has two vessels which allow maintaining the operation when one vessel
is off for maintenance purposes, as illustrated in Figure 9.17. Also, option 2 is estimated to increase the revenue per year by $ 200,000.

Considering the data in Table 9.9, what is the best option from a cost perspective?

9.6 CONTRACT COST UPDATES – READJUSTMENT

Contract cost updates occur when the duration is longer than 1 year and when inflation is real. It is an example of a parametric equation to update the cost. Hence, the
equation helps protect the contract from inflation and escalation effects.

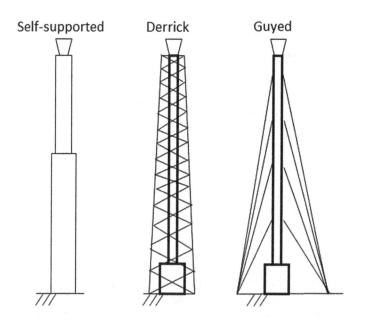

FIGURE 9.16   Flare structure types

TABLE 9.8
Value Engineering Example

Description	Derrick	Guyed	Self supported
CAPEX ($ million)	35	28	33
OPEX ($ million/year)	1	1.5	1.2
Operation life (years)	25	25	25
Total ($ million)	$35 + 25 \times 1 = 60$	$28 + 25 \times 1.5 = 66$	$33 + 25 \times 1.2 = 63$

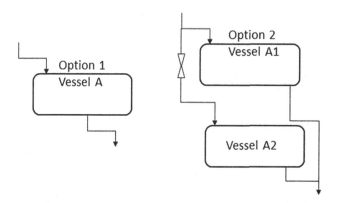

FIGURE 9.17   Value engineering – Exercise 4, options schematic

TABLE 9.9

Value Engineering – Exercise 4

Description	Option 1	Option 2
CAPEX ($ million)	2	3.4
OPEX ($ million/year)	0.1	0.15
Operation life (years)	20	20
Revenue increment ($ million)	-	0.2

FIGURE 9.18   Cost updates in contracts

Figure 9.18 shows the sequence of creating the readjustment.

The cost categorization breaks the contract scope into categories, such as labor, material, and equipment. Then, each category should have a specific economic index, and the selection should consider factors like location and data availability. For example, a labor index from the US does not apply to Africa. Also, the availability of indices is not the same. Some countries have a massive variety of indices, and others do not. In this case, generic indices could be analyzed.

Equations 9.6 and 9.7 show a generic example of the parametric equation. The equation could be different for each scope. For example, the labor factor could be removed if it is not relevant. Also, there could be more factors and other indexes for material types, such as raw, bulk, and fabricated material.

$$C_R = C_0 \times \left(f_a \times \frac{LI_1}{LI_0} + f_b \times \frac{MI_1}{MI_0} + f_c \times \frac{EI_1}{EI_0} + f_d \times \frac{AI_1}{AI_0} \right) \quad \text{(Equation 9.6)}$$

$$f_a + f_b + f_c + f_d = 1 \quad \text{(Equation 9.7)}$$

Where:

C_r = Readjusted cost
C_0 = Original cost
0 = Index on the original date
1 = Index on the base date (desired)
LI = Labor index
MI = Material index
EI = Equipment index
AI = Additional cost index
fn = A percentage which represents each category's costs about the total cost

The example includes adjusting a service contract from January 2022 to December 2023. Table 9.10 provides the contract breakdown when it was awarded in January 2022, and Table 9.11 shows the index value per month (only a few months are shown

TABLE 9.10
Contract Cost Breakdown

Category	Value ($) Jan 2022	f (%)
Total ($)	512,000	100
Labor ($)	200,000	39.1
Material ($)	120,000	23.4
Equipment ($)	140,000	27.3
Additional costs ($)	52,000	10.2

TABLE 9.11
Economic Indexes

Period	Labor index (LI)	Material index (MI)	Equipment index (EI)	Other costs index (AI)
Jan-22	7,200	240.9	217.7	249.4
Feb-22	7,250		219.2	249.9
Mar-22	7,290		220.1	251.1
...	...	...	...	...
Oct-23	7,510	249.9	234.1	257.1
Nov-23	7,590	251.5	234.9	257.6
Dec-23	7,646	252.4	236.4	258.4

for simplicity). Also, the indicators' names are generic just for illustration purposes, and the numbers of each indicators' numbers are hypotheticals.

Through the Equation 9.6, the result is:

$$C_R = 512,000 \times \left(0.391 \times \left(\frac{7646}{7200} \right) + 0.234 \times \left(\frac{252.4}{240.9} \right) \right.$$

$$\left. +0.273 \times \left(\frac{236.4}{217.7} \right) + 0.102 \times \left(\frac{258.4}{249.4} \right) \right) =$$

$$C_R = \$ 544,019 \text{ or } + 6.3\%$$

The result shows that the value for service increased by 6% during the period.

9.6.1 Contract Cost Updates – Exercise 5

Company ABC wants to calculate the contract cost in December 2023 of the contract awarded in January 2021. The contract cost breakdown is shown in Table 9.12, and economic indexes are in Table 9.13. What category has the highest absolute and/or percentage cost increase in the period?

TABLE 9.12

Contract Cost Breakdown – Exercise 5

Category	Value ($) Jan 2021	f (%)
Total ($)	1,000,000	100
Labor ($)	300,000	30
Material ($)	300,000	30
Equipment ($)	150,000	15
Construction equipment ($)	150,000	15
Additional costs ($)	100,000	10

TABLE 9.13

Economic Indexes – Exercise 5

Period	Labor index (LI)	Material index (MI)	Equipment index (EI)	Construction equipment (CEI)	Other costs index (AI)
Jan-21	5,000	100	217.7	310	249.4
Feb-21	5,010	101	219.2	333	249.9
Mar-21	5,020	102	220.1	350	251.1
...	...	...	...		...
Oct-23	5,095	115	239.3	400	260.1
Nov-23	5,110	116	241.2	425	263.4
Dec-23	5,120	117	243.9	450	269.5

NOTES

1. Reprinted with permission from AACE International. Check the website for the latest versions (https://web.aacei.org/resources/cost-engineering-terminology).
2. Reprinted with permission from AACE International. Check the website for the latest versions (https://web.aacei.org/resources/cost-engineering-terminology).
3. Reprinted with permission from AACE International. Check the website for the latest versions (https://web.aacei.org/resources/cost-engineering-terminology).

BIBLIOGRAPHY

AACE International, 2022. *Recommended Practice 10S-90 Cost Engineering Terminology.* Available at: https://web.aacei.org/resources/cost-engineering-terminology. [Accessed: 8 January 2023].

BLS, 2015. *Handbook of Methods.* Available at: https://www.bls.gov/opub/hom/home.htm. [Accessed: 7 May 2023].

Department of Defense, 2016. *Inflation and Escalation Best Practises for Cost Analysis.* Available at: https://www.cape.osd.mil/files/InflationandEscalationBestPracticesforCostAnalysisforWebsiteForPubRelReview.pdf. [Accessed: 7 May 2023].

GAO, 2020. Cost Estimating and Assessment Guide. Available at: https://www.gao.gov/prod-ucts/gao-20-195g. [Accessed: 15 December 2022].

Hastak, M., 2015. *Skills & Knowledge of Cost Engineering*, 6th Edition. AACE International.

Macedo, H., 2014. *Cost Updates in Global Contracts*. AACE International.

Oxford, 2016. *Dictionary of Business and Management*, 6th Edition. Oxford University Press.

RICS, 2017. *Value Management and Value Engineering*, 1st Edition. RICS Professional Standard and Guidance, UK. Available at: https://www.rics.org/profession-standards /rics-standards-and-guidance/sector-standards/construction-standards/black-book/ value-management-and-value-engineering-1st-edition. [Accessed: 28 April 2023].

Tennent, J., 2013. *Guide to Financial Management*, 2nd Edition. The Economist in Association with Profile Books Ltd.

EXERCISE ANSWER

1- Figure "Exercise Answer" shows the three dots, which show projects A, B, and C. Project A has the same result as the industry benchmark, which is excellent. Project B is above the benchmarking but under the upper limit, which is a satisfactory result. Project C must be reviewed or cancelled because it exceeds the benchmark and the upper limit.

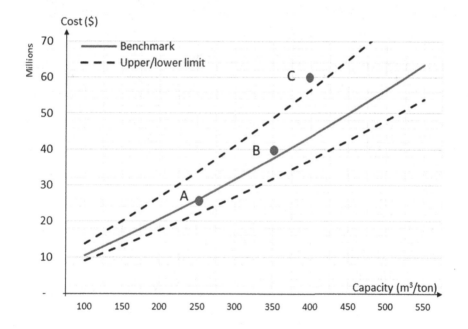

2- See table below

Year	Total depreciation	Asset 1	Asset 2	Asset 3	Asset 4
1	4,584,615	450,000	2,000,000	1,384,615	750,000
2	4,619,231	450,000	1,840,000	1,329,231	1,000,000
3	4,916,646	450,000	1,692,800	1,273,846	1,500,000
4	5,100,838	450,000	1,557,376	1,218,462	1,875,000
5	5,170,863	450,000	1,432,786	1,163,077	2,125,000
6	5,375,855	450,000	1,318,163	1,107,692	2,500,000
7	5,090,018	450,000	1,212,710	1,052,308	2,375,000
8	4,687,616	450,000	1,115,693	996,923	2,125,000
9	4,292,976	450,000	1,026,438	941,538	1,875,000
10	3,780,477	450,000	944,323	886,154	1,500,000
11	3,399,546	450,000	868,777	830,769	1,250,000
12	3,024,659	450,000	799,275	775,385	1,000,000
13	2,655,333	450,000	735,333	720,000	750,000
14	2,416,122	450,000	676,506	664,615	625,000
15	2,306,616	450,000	622,386	609,231	625,000
16	2,201,441	450,000	572,595	553,846	625,000
17	2,100,249	450,000	526,787	498,462	625,000
18	2,002,721	450,000	484,644	443,077	625,000
19	1,908,565	450,000	445,873	387,692	625,000
20	1,534,844	450,000	127,536	332,308	625,000
21	276,923		-	276,923	
22	221,538		-	221,538	
23	166,154		-	166,154	
24	110,769		-	110,769	
25	55,385		-	55,385	

3- Original Price from Chapter 4 – $ 2,345,770

  Factor 1 – Tower cost reviewed. Result $ 2,480,664. Increased in + $ 134.8k

  Factor 2 – The benefits are revised. Result $ 2,275,239. Reduction – $ 70.5k

  Factor 3 – The lift services are subcontracted, costing $ 30,000 monthly. Result $ 2,295,771. Reduction – $ 49.9k

  Factor 4 – Interest rate increases to 5% and tax to 25%. Result $ 2,602,733. Increased in + $ 256.9k

  Factor 4 has the highest negative impact, and factor 2 has the positive one.

4- Option 1 – $ 4.0 million. Option 2 – $ 2.4 million. Option 2 is the best.

5- Contract readjusted: $ 1,152,054 or + 15.2%

10 Sustainability and Cost

The search for a sustainable world should engage all science areas, and cost engineering should be an actor in addressing carbon emission reduction actions. The section shows basic concepts and how to estimate carbon emissions during the asset lifecycle. It aims to create scenarios to analyze solutions that combine cost and carbon emission reductions.

Several concepts are necessary to measure and analyze the carbon emissions:

Carbon footprint is *"the total amount of CO2 and other greenhouse gases, emitted over the full life cycle of a process or product"*

(UK Parliamentary, 2006).

Embodied carbon is the carbon arising from construction or maintenance and disposal at the end of the life of an asset.

Cap-and-trade system is a system that defines a total amount (limit) of greenhouses gases that members can emit during a period (e.g., a year). These members can buy or sell emissions allowances. One example, is the European Trading Scheme.

10.1 EMBODIED CARBON CALCULATION

The embodied carbon calculation is divided into several steps, as shown in Figure 10.1. The sequence is the material selection, source's definition, boundary, transportation, waste disposal, calculation, and analysis.

Embodied carbon calculation should be done for each material of the asset or through pareto principle, for the most relevant ones which represent the majority of the carbon emissions.

Each material used during the construction, decommissioning, and maintenance could be used. Then source's definition aims to define a metric of carbon per unit of measure for the material selected from a confident source. Several examples are:

- Inventory of Carbon and Energy (ICE)
- Environmental Performance in Construction (EPiC)
- Department for Business, Energy, and Industrial Strategy – UK Government
- Environmental Product Declarations (EPDs)

Then, the metric should inform the boundary conditions. There are four options shown in Figure 10.2. Cradle to gate means it covers only the embodied carbon during manufacturing. Cradle to the site includes transportation, and cradle to grave covers all the life until the waste disposal. Finally, there is cradle to cradle when the material is reused in a second life.

DOI: 10.1201/9781003402725-10

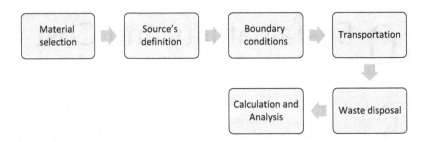

FIGURE 10.1 Carbon footprint sequence

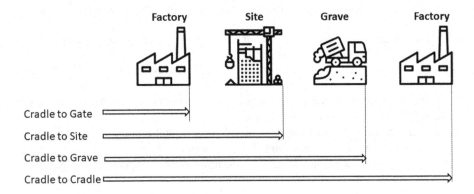

FIGURE 10.2 Boundary conditions and icons made by Freepick from www.flaticon.com

When the boundary conditions are cradle to gate or to the site, the embodied carbon during the transportation and waste disposal should be estimated and added. Finally, all components of embodied carbon are added and summarized, and the analysis is done to check the best solution that combines low cost and low emission.

10.1.1 EMBODIED CARBON CALCULATION – EXAMPLE 1

The example is based on the analyses of two material types for insulation: Rockwool and Expanded Polystyrene insulation (EPS). These materials are selected because they have properties to work as insulation. EPS provides low thermal conductivity value, high thermal insulation, high mechanical compressive strength, repels water, and prevents humidity and bacterial growth. Rockwool is a vapor-permeable material. Its insulation value (U-value) remains unchanged over time. It is a suitable material to prevent noise and fire, and it does not cause that much skin irritation.

The source adopted is ICE 2.0, 2011. Table 10.1 lists the embodied carbon value for each material. Also, the ICE inform us that the boundary conditions are cradle to grave. Hence, no estimation for transportation and waste disposal is necessary.

The assumption is an area of 100 m² and 150 mm of thickness should be isolated. Also, the material waste of the material is 10%, and the density is 60 kg/m³ for

TABLE 10.1
Embodied Carbon per Material, Table Adapted from ICE 2.0, 2011

Material	Embodied carbon
Rockwool	1.12 kgCO$_2$/Kg
EPS	3.29 kgCO$_2$/Kg

TABLE 10.2
Embodied Carbon Calculation per Material

Material	Volume (m³)	Mass (kg)	Waste (10%)	Embodied carbon	Total (tCO2e)
Rockwool	100 m² × 150 mm = 15 m³	15 × 60 kg/m³ = 900 kg	90 kg	1.12 kgCO$_2$/Kg	1.1
EPS	100 m² × 150 mm = 15 m³	15 × 30 kg/m³ = 450 kg	45 kg	3.29 kgCO$_2$/Kg	1.6

Rockwool and 30 kg/m³ to the EPS. The analysis could be done through this information, as shown in Table 10.2.

The total is calculated below:

Rockwool embodied carbon = (900 kg + 90 kg) × 1.12 kgCO2/kg = 1.1 tCO2e
EPS embodied carbon = (450 kg + 45 kg) × 3.29 kgCO2/kg = 1.6 tCO2e

Hence, the best option for carbon embodied perspective is the Rockwool.

10.1.1 Scenario Analysis – Example 2

Energy Company ABC has two possible solutions for the new process plant. Option A has a CAPEX of $ 10 million and embodied carbon calculation estimated is 150,000 tCO2e. Option B has a higher CAPEX, 11 million, but a lower embodied carbon estimation, 120,000 tCO22.

Considering that tCO2e can be sold for $ 50, what is the best option?

The CO2e difference between the two options are:

150,000 − 120,000 = 30,000 tCO2e.
Selling the difference: 30,000 × $ 50 = $ 1.5 million

Hence, option B is more rentable considering this scenario:

10,000,0000 > 11,000,000 − 1,500,000 = 9,500,000

Several factors should be considered in real-life scenarios, but the example aims to show that one cap-and-trade system can include additional aspects during the economic feasibility study. Also, it can improve their image to the customers.

10.2 DECOMMISSIONING

Decommissioning is the process to deinstall the asset. It can be done by reuse, recycling, storage, and/or disposal the materials and equipment.

There is a direct correlation between decommissioning and sustainability because the correct destination can mitigate environmental disaster as leakages and promote a second life through recycling or reuse techniques.

Reuse aims to create a second life for each asset component through repairs, refurbishment, or remanufacturing. Recycling consists of converting equipment or material into reusable material.

10.2.1 DECOMMISSIONING – EXAMPLE 3

Considering Example 3 in Section 3.2.1 and the data below, how much will the decommissioning of the solar farm for the two scenarios below:

Solar Farm – Scenario 1 – Disposal
  Disposal cost per panel = $ 10
  Environmental fee = $ 500,000 per adopted the disposal solution.
  Number of panels = 40,000 PV panels
Solar Farm – Scenario 2 – Reuse and Recycling
  Recycling cost per panel = $ 20
  Assumption 20,000 PV panels will be recycled
  Reuse cost per panel (transportation and installation) = $ 15
  Assumption 20,000 will be reused to support poor local communities.

Analysis

  Scenario 1 = Total cost = $ 500,000 + 40,000 × 10 = $ 900,000
  Scenario 2 = Total cost = $ 20 × 20,000 + $ 15 × 20,000 = $ 700,000

Scenario 2 becomes economically feasible when the analysis considers the environmental fee.

10.2.2 DECOMMISSIONING – EXERCISE 1

Energy Company ABC has one asset that should be decommissioned. What is the best method for each WBS element considering the data listed in table 10.3?

Assumptions: One method should be selected per line of the WBS for simplification. N/A means that the method does not apply to that specific scope.

TABLE 10.3

Decommissioning Exercise 1 – Cost per Method and WBS Element

ID	WBS	Disposal cost ($)	Reuse cost ($)	Recycling cost ($)
1	Containers offices	15,000	14,000	20,000
2	Warehouse	50,000	N/A	45,000
3	Utilities facilities			
3.1	Piping	30,000	70,000	20,000
3.2	Civil components (e.g. concrete supports)	25,000	50,000	22,000
3.3	Steel structures	40,000	60,000	42,000
3.4	Equipments	60,000	50,000	90,000
3.5	Electrical (e.g., cables, modules)	12,000	N/A	9,000
3.6	Instruments	6,000	8,000	10,000
4	Roads and access	60,000	N/A	70,000
5	Process plant			
5.1	Piping	90,000	120,000	70,000
5.2	Equipments	120,000	100,000	110,000
5.3	Civil components (e.g. concrete supports)	30,000	N/A	35,000
5.4	Electrical (e.g., cables, modules)	34,000	N/A	29,000
5.5	Steel structures	85,000	100,000	75,000
5.6	Instruments	9,000	8,000	12,000

The disposal cost includes transportation, uninstall, and associated fees.

Reuse cost considers repairs (refurbishment), uninstall, transportation, and installation costs minus the selling (revenue) profit for the second life.

Recycling cost includes the removal, transportation, and recycling costs minus the selling (revenue) of the recycled material.

BIBLIOGRAPHY

Building Transparency, no date. *EC3 User Guide*. Available at: https://www.buildingtra nsparency.org/ec3-resources/ec3-user-guide/. [Accessed: 13 May 2023].

Business, Energy and Industrial Strategy – UK Government, 2019. *Decommissioning of Offshore Renewable Energy Installations under the Energy Act 2004 – Annex C.* Available at: https://assets.publishing.service.gov.uk/government/uploads/system/ uploads/attachment_data/file/916912/decommisioning-offshore-renewable-energy -installations-energy-act-2004-guidance-industry__1_.pdf. [Accessed: 18 August 2023].

Circular Ecology, no date. *Embodied Carbon – The ICE Database*. Available at: www.circu- larecology.com/ice-database.html. [Accessed: 7 May 2023].

Department For Energy Security, 2022. *Greenhouse Gas Reporting: Conversion Factors 2022*. Available at: https://www.gov.uk/government/publications/greenhouse-gas -reporting-conversion-factors-2022. [Accessed: 13 May 2023].

TABLE 10.4

Decommissioning Selected Method per WBS Element

ID	Description	Disposal cost ($)	Reuse cost ($)	Recycling cost ($)
1	Containers offices	No	Yes	No
2	Warehouse	No	No	Yes
3	Utilities facilities			
3.1	Piping	No	No	Yes
3.2	Civil components (e.g., concrete supports)	No	No	Yes
3.3	Steel structures	Yes	No	No
3.4	Equipments	No	Yes	No
3.5	Electrical (e.g., cables, modules)	No	No	Yes
3.6	Instruments	Yes	No	No
4	Roads and access	Yes	No	No
5	Process Plant			
5.1	Piping	No	No	Yes
5.2	Equipments	No	No	Yes
5.3	Civil components (e.g., concrete supports)	Yes	No	No
5.4	Electrical (e.g., cables, modules)	No	No	Yes
5.5	Steel structures	No	No	Yes
5.6	Instruments	No	Yes	No

European Commission, no date. *EU Emissions Trading System*. Available at: https://climate
.ec.europa.eu/eu-action/eu-emissions-trading-system-eu-ets_en#a-cap-and-trade-sys-
tem. [Accessed: 12 May 2023].

Invernizzi et al., 2020. Developing policies for the end-of-life of energy infrastructure:
Coming to terms with the challenges of decommissioning. *Journal Energy Policy*, 144,
111677. doi.org/10.1016/j.enpol.2020.111677.

Melbourne School of Design, no date. *EPiC Database*. Available at: https://msd.unimelb.edu
.au/research/projects/current/environmental-performance-in-construction/epic-data-
base. [Accessed: 13 May 2023].

Solar Power World, 2019. *Old Solar Panels Get Second Life in Repurposing and Recycling
Markets*. Available at: https://www.solarpowerworldonline.com/2019/01/old-solar-pan-
els-get-second-life-in-repurposing-and-recycling-markets/. [Accessed: 3 April 2021].

EXERCISE ANSWER

1- Decommissioning total cost $ 588,000. Disposal – $ 136,000; Reuse – $
72,000; Recycling – $ 380,000. The result is based on the selection accord-
ing to Table 10.4.

Annex

ANNEX I: STATISTICS AND UNITS OF MEASUREMENT

Seven concepts are listed through a definition, formula, and examples based on Table I.1. They are Mean, Median, Mode, Range, Variance, Standard Deviation, and Percentile. After that, Normal Distribution, Cumulative Frequency Curve, and Significant figures are illustrated.

1) Mean (μ)

Mean (average) is the sum of measurements divided by the number of measurements.

$$\mu = \sum x / N \qquad \text{(Equation I.1)}$$

Where:
N = number of sample
X = each of element of sample
For example, the <u>Age</u> mean is calculated below:

$$\text{Age } \mu = \frac{(19 + 21 + 25 + 27 + 28 + 33 + 33 + 33 + 37 + 38 + 40 + 40 + 48 + 50 + 50)}{15 \, \text{professionals}}$$

$$= \frac{523}{15}$$

$$\boldsymbol{\mu = 34.8}$$

Repeating the calculation for the <u>Salary</u> mean:

$$\mu = (2010 + 1600 + 2600 + 3000 + 2000 + 2500 + 3000 + 3600 + 4000 + 2600 + 3100 + 2600 + 4100 + 3500 + 2900) / 15 \text{ professionals}$$

$$\mu = 43{,}110/15$$

$$\mu = \boldsymbol{2{,}874}$$

2) Median ($\bar{x}$)

It is the middle number when the data observations are arranged in ascending or descending order.

TABLE I.1

Data for Examples in the Statistics Concepts

Name	Age	Monthly Salary ($)
1. Paul	19	2,010
2. Joseph	21	1,600
3. John	25	2,600
4. Carl	27	3,000
5. Simon	28	2,000
6. Bob	33	2,500
7. Zak	33	3,000
8. Mike	33	3,600
9. Lis	37	4,000
10. Joe	38	2,600
11. Rick	40	3,100
12. Marco	40	2,600
13. Mary	48	4,100
14. Ana	50	3,500
15. Negan	50	2,900
Total	**523**	**43,110**

For example, the <u>Age</u> Median is shown below:

$$19, 21, 25, 27, 28, 33, 33, \boxed{33}, 37, 38, 40, 40, 48, 50, 50$$

Age Median ($\bar{x}$) is 33, because it is in the middle of the data or eighth position.
   For example, the <u>Salary</u> Median is shown below:

$$1600, 2000, 2010, 2500, 2600, 2600, 2600, \boxed{2900}, 3000, 3000, 3100, 3500, 3600,$$

$$4000, 4100$$

Salary Median ($\bar{x}$) is 2,900, because it is in the middle of the data or eighth position.

3) Mode

Mode is the measurement that occurs most often in the data set.
   For example, the <u>Age</u> Mode is shown below:

$$19, 21, 25, 27, 28, \boxed{33, 33, 33}, 37, 38, 40, 40, 48, 50, 50$$

Age Mode is 33 because it is the highest number of repetitions, 3.

For example, the <u>Salary</u> Mode is shown below:

$$1600, 2000, 2000, 2500, \boxed{2600, 2600, 2600}, 2900, 3000, 3000, 3100, 3500,$$

$$3600, 4000, 4100$$

Salary Mode is 2,600 because it is the highest number of repetitions, 3.

4) **Range**

Range is the difference between the data set's largest and smallest values. It is calculated by Equation I.2.

$$\text{Range} = \text{largest value} - \text{smallest value} \qquad \text{(Equation I.2)}$$

For example, the <u>Age</u> Range is shown below:

$$\boxed{19}, 21, 25, 27, 28, 33, 33, 33, 37, 38, 40, 40, 48, 50, \boxed{50}$$

$$\textbf{Range} = 50 - 19 = \textbf{31}$$

For example, the <u>Salary</u> Range is shown below:

$$\boxed{1600}, 2000, 2010, 2500, 2600, 2600, 2600, 2900, 3000, 3000, 3100,$$

$$3500, 3600, 4000, \boxed{4100}$$

$$\textbf{Range} = 4100 - 1600 = \textbf{2500}$$

5) **Variance (σ^2)**

Variance is the average of the squared deviations from the mean.
Population variance is estimated by Equation I.3:

$$\sigma^2 = \sum_{i=1}^{N} (x_i - \mu)^{\wedge}2 \, / N \qquad \text{(Equation I.3)}$$

Where:
μ = Mean
x_i = Each of the element of sample
N = Population
When a sample is available, not the population, Equation I.4 should be used:

$$s^2 = \sum_{i=1}^{n} (x_i - \mu)^{\wedge}2 \, / (n - 1) \qquad \text{(Equation I.4)}$$

Where:

μ = Mean

x_i = Each of element of sample

n = Total number of elements

For example, the Age Variance is shown in Table I.2. In this case, all are considered, so the population equation is used. The mean is 34.8.

Then the Variance is:

$$\sigma^2 = 1327.8 / 15 = \mathbf{88.5}$$

For example, the <u>Salary</u> Variance is shown in Table I.3. This case all are considered, so population equation is used. The mean is 2,874.

Then the Variance is:

$$\sigma^2 = 7,271,960 / 15 = \mathbf{484,797.3}$$

6) Standard Deviation (σ)

Standard Deviation is the positive square root of the variance. It is estimated by Equation I.5:

$$\sigma = (Variance)^{0.5} \qquad \text{(Equation I.5)}$$

TABLE I.2
Variance Calculation for Age

Name	Age	$(x_i - \mu)^2$
Paul	20	$(20 - 34.8)^2 = 219.04$
Joseph	21	$(21 - 34.8)^2 = 190.44$
John	25	$(25 - 34.8)^2 = 96.04$
Carl	27	$(27 - 34.8)^2 = 60.84$
Simon	28	$(28 - 34.8)^2 = 46.24$
Bob	33	$(33 - 34.8)^2 = 3.24$
Zak	33	$(33 - 34.8)^2 = 3.24$
Mike	33	$(33 - 34.8)^2 = 3.24$
Lis	37	$(37 - 34.8)^2 = 4.84$
Joe	38	$(38 - 34.8)^2 = 10.24$
Rick	40	$(40 - 34.8)^2 = 27.04$
Marco	40	$(40 - 34.8)^2 = 27.04$
Mary	48	$(48 - 34.8)^2 = 174.24$
Ana	50	$(50 - 34.8)^2 = 231.04$
Negan	50	$(50 - 34.8)^2 = 231.04$
Sum	523	1,327.8

TABLE I.3
Variance Calculation for Salary

Name	Salary	$(x_i - \mu)^2$
Paul	2,010	$(2{,}010 - 2{,}874)^2 = 746{,}496$
Joseph	1,600	$(1{,}600 - 2{,}874)^2 = 1{,}623{,}076$
John	2,600	$(2{,}600 - 2{,}874)^2 = 75{,}076$
Carl	3,000	$(3{,}000 - 2{,}874)^2 = 15{,}876$
Simon	2,000	$(2{,}000 - 2{,}874)^2 = 763{,}846$
Bob	2,500	$(2{,}500 - 2{,}874)^2 = 139{,}876$
Zak	3,000	$(3{,}000 - 2{,}874)^2 = 15{,}876$
Mike	3,600	$(3{,}600 - 2{,}874)^2 = 527{,}076$
Lis	4,000	$(4{,}000 - 2{,}874)^2 = 1{,}267{,}876$
Joe	2,600	$(2{,}600 - 2{,}874)^2 = 75{,}076$
Rick	3,100	$(3{,}100 - 2{,}874)^2 = 51{,}076$
Marco	2,600	$(2{,}600 - 2{,}874)^2 = 75{,}076$
Mary	4,100	$(4{,}100 - 2{,}874)^2 = 1{,}503{,}076$
Ana	3,500	$(3{,}500 - 2{,}874)^2 = 391{,}876$
Negan	2,900	$(2{,}900 - 2{,}874)^2 = 676$
Sum	43,110	7,271,960

For example, the <u>Age</u> Standard Deviation Variance is shown below.

$$\sigma = (88.5)^{0.5} = 9.4$$

And the <u>Salary</u> Standard Deviation

$$\sigma = (484{,}797.3)^{0.5} = 696.3$$

7) Percentile

Percentile is the number with exactly p percent of the measurements fall below it and $(100-p)$ per cent fall above it when the data are arranged in ascending or descending order. There are three famous percentiles:

- The first (lower) quartile: the 25th percentile
- The second (middle) quartile: is the median.
- The third (upper) quartile: the 75th percentile.

For example, the percentiles for age and salary from Table I.1 are shown below.

Table I.1 has 15 numbers. Hence, the 25% of population is 3.75 ($15 \times 25\%$) or round up the 4th position, the 50% is 7.5 ($15 \times 50\%$) or round up the 8th position, and the 75% is 11.25 ($15 \times 75\%$) or round down the 11th position.

$$20, 21, 25, \boxed{27}, 28, 33, 33, \boxed{33}, 37, 38, 40, \boxed{40}, 48, 50, 50$$

First   Second(Median)   Third

$$1600, 2000, 2000, \boxed{2500}, 2600, 2600, 2600, \boxed{2900}, 3000, 3000, 3100, \boxed{3500},$$

First   Second(Median)   Third

$$3600, 4000, 4100$$

8) Normal Distribution

Normal Distribution or Gaussian Distribution is "the symmetrical bell-shaped frequency curve formed when the frequency of arrange of values is plotted on the vertical axis against the value of a random variable on the horizontal axis (Oxford, 2016)."

It is calculated by Equation I.6.

$$f(x) = \frac{1}{\sigma\sqrt{2\pi}} e^{\frac{-1}{2}\frac{(x-\mu)^2}{\sigma^2}}$$

(Equation I.6)

Where

σ = standard variation

μ = mean

The Normal Distribution curve is shown in figure I.1. Both sides are symmetrical, and it has an asymptotic tail. Also, the median, mean, and mode have the same result.

9) Cumulative Frequency Curve

The Cumulative Frequency Curve represents the cumulative frequency distribution of grouped data on a graph. Table I.4 lists a group of data in column A, column B shows the cumulative frequency of the data, and column C shows the percentage vision.

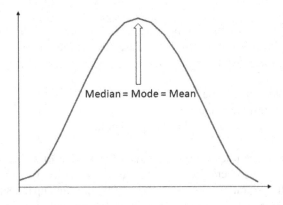

FIGURE I.1 Normal distribution curve.

TABLE I.4
Cumulative Data Example

ID	A	B	C
Horizontal Axis	Data	Cumulative Frequency Data	Cumulative Percentage (%)
0	0	0	0
1	20	0 + 20 = 20	2
2	72	20 + 72 = 92	9
3	122	92 + 122 = 214	21
4	155	214 + 155 = 369	37
5	164	369 + 164 = 533	53
6	150	533 + 150 = 683	68
7	121	683 + 121 = 804	80
8	87	804 + 87 = 891	89
9	55	891 + 55 = 946	95
10	32	946 + 32 = 978	98
11	16	978 + 16 = 994	99
12	6	994 + 6 = 1000	100

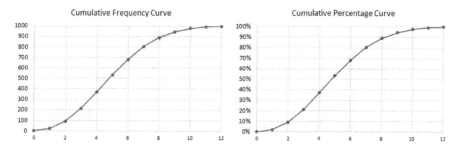

FIGURE I.2 Cumulative frequency curve (left) and cumulative percentage curve (Right).

Figure I.2 shows both the Cumulative Frequency Curve, left, and the Cumulative Percentage Curve, right, from data in Table I.4.

UNITS OF MEASUREMENT

Generally, for basic topics, the correct attention is not given. Unit measurement is basic knowledge, but a mistake can create a big problem or lose a business/tender.

It is important to create mechanisms or procedures to check the math and units measurement and to ensure the correct information.

Multiplying factor is listed in Table I.5. For example, if a manager says that the pump cost is $ 5k, it means the cost is 5000 dollars.

Figure I.3 shows the conversions for mass (a), time (b), length (c), and volume (d). For example, tonne is converted by the kilogram, multiplied by 1000, and kilogram

TABLE I.5
Multiplying Factor

Prefix	Symbol	Multiplying Factor
Giga	G	$10^9 = 1\ 000\ 000\ 000$
Mega	M	$10^6 = 1\ 000\ 000$
Kilo	k	$10^3 = 1\ 000$
Milli	m	$10^{-3} = 0.001$
Micro	μ	$10^{-6} = 0.000\ 001$

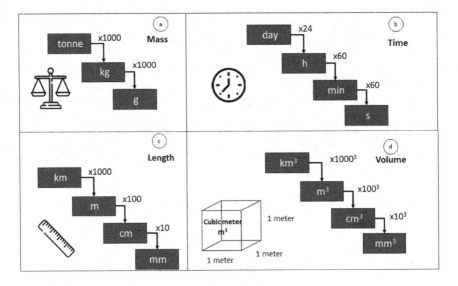

FIGURE I.3   Units of measurement: Mass (a), Time (b), Length (c), and Volume (d).

to the gram, times one thousand as well. The time factor between second, minute, and hour is 60, but the converter between day and hour is 24.

The length converter factor starts with 1,000 from kilometer to meter, then 100 to converter to the centimeter, and finally, 10 should be multiplied to achieve the millimeter. The volume conversion is similar, but each factor is potential by 3 because it is a cubic conversion.

Figure I.4 shows the mass and length and how to convert the metric system to the Imperial/US system and vice versa. For example, 1 km is 0.621 miles, or one mile is 1.61 km.

SIGNIFICANT FIGURES

They are the number of digits that contribute to the accuracy of it.

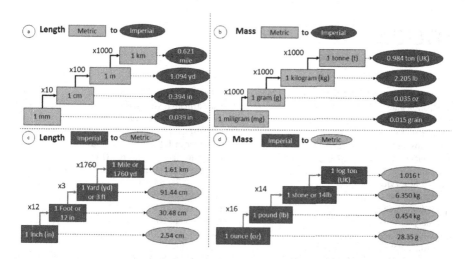

FIGURE I.4   Convertion the metric system to the imperial/US system and vice versa.

Rules

Zeros to the left of the first non-zero digit are NOT significant.

0.15 → 2 significant
0.0031 → 2 significant

Zeros between the significant non-zero digits ARE significant.

201.5 → 4 significant
1.06 → 3 significant

Numbers between 1 and 9 are significant.
Addition and Subtraction
The result must respect the number with lowest significant figures.

9.5 + 10.12 = 19.62 = 20
320.3 − 220.10 = 100.20 = 100.2

Multiplication and Division
The result must respect the number with the lowest significant figures.

3.5 × 2.33 = 8.155 = 8.2
6.20/2 = 3.1 = 3

BIBLIOGRAPHY

Hastak, M., 2015. *Skills & Knowledge of Cost Engineering*, 6th Edition. AACE International.
Oxford, 2016. *Dictionary of Business and Management*. 6th Edition. Oxford University Press.

Index

Printed in the United States
by Baker & Taylor Publisher Services